AF829226

Oil Export Economies:
New Comparative Perspectives
on the Arab Gulf States

Other Books on the Gulf Region Published by Gerlach Press:

The GCC in the Global Economy
Ed. by Richard Youngs
ISBN 9783940924018, 2012

Resources Blessed: Diversification and the Gulf Development Model
Ed. by Giacomo Luciani
ISBN 9783940924025, 2012

GCC Financial Markets: The World's New Money Centers
Ed. by Eckart Woertz
ISBN 9783940924032, 2012

National Employment, Migration and Education in the GCC
Ed. by Steffen Hertog
ISBN 9783940924049, 2012

Asia-Gulf Economic Relations in the 21st Century – The Local to Global Transformation
Ed. by Tim Niblock with Monica Malik
ISBN 9783940924100, 2013

A New Gulf Security Architecture – Prospects and Challenges for an Asian Role
Ed. by Ranjit Gupta, Abubaker Bagader, Talmiz Ahmad and N. Janardhan
ISBN 9783940924360, 2014

Gulf Charities and Islamic Philanthropy in the 'Age of Terror' and Beyond
Ed. by Robert Lacey and Jonathan Benthall
ISBN 9783940924322, 2014

The Silent Revolution – The Arab Spring and the Gulf States
Ed. by May Seikaly and Khawla Mattar
ISBN 9783940924346, 2014

Islamic Finance – Political Economy, Performance and Risk
Ed. by Mehmet Asutay and Abdullah Turkistani
ISBN 9783940924124, 3 vols set, 2015

www.gerlach-press.de

Oil Export Economies: New Comparative Perspectives on the Arab Gulf States
Annika Kropf
ISBN 9783940924582, 2016

Employment and Career Motivation in the Arab Gulf States: The Rentier Mentality Revisited
Ed. by Annika Kropf and Mohamed Ramady
ISBN 9783940924605, 2015

The Changing Energy Landscape in the Gulf: Strategic Implications
Ed. by Gawdat Bahgat
ISBN 9783940924643, 2015

Sustainable Development Challenges in the Arab States of the Gulf
Ed. by David Bryde, Yusra Mouzughi and Turki Al Rasheed
ISBN 9783940924629, 2015

Africa and the Gulf Region: Blurred Boundaries and Shifting Ties
Ed. by Rogaia Mustafa Abusharaf and Dale F. Eickelman
ISBN 9783940924704, 2015

Reconstructing Yemen: Governance, Economy, and Society
Ed. by Noel Brehony and Saud Al-Sarhan
ISBN 9783940924681, 2015

State-Society Relations in the Arab Gulf States
Ed. by Mazhar Al-Zoby and Birol Baskan
ISBN 9783940924384, 2014

Political Economy of Energy Reform – The Clean Energy-Fossil Fuel Balance in the Gulf
Ed. by Giacomo Luciani and Rabia Ferroukhi
ISBN 9783940924407, 2014

The Politics of Food Security Asian and Middle Eastern Strategies
Ed. by Sara Bazoobandi
ISBN 9783940924308, 2014

Annika Kropf

Oil Export Economies: New Comparative Perspectives on the Arab Gulf States

With a Foreword by Giacomo Luciani

Gerlach Press

First published 2016
by Gerlach Press
Berlin, Germany
www.gerlach-press.de

Cover Design: www.brandnewdesign.de, Hamburg
Printed and bound in Germany by Hubert & Co, Göttingen
www.hubertundco.de

© Gerlach Press 2016
All rights reserved. No part of this publication may be reprinted or reproduced, or utilised in any form or by any electronic, mechanical, or other means, now known or hereafter invented, including photocopying and recording, or in any other information storage or retrieval system, without permission in writing from the publisher.

British Library Cataloguing in Publication Data.
A catalogue record for this book is available from the British Library.

Bibliographic data available from Deutsche Nationalbibliothek
http://d-nb.info/1068029293

ISBN: 978-3-940924-58-2 (hardcover)
ISBN: 978-3-940924-59-9 (ebook)

Contents

Foreword

Annika Kropf's book on the Arab Gulf States' Oil Export Economies sees the light at a crucial time in the economic history and development trajectory of the six GCC countries. Their process of economic diversification has reached a crucial crossroad: faced with the sudden and unexpected collapse in oil prices, the time has come for them to demonstrate that progress has indeed been achieved towards reduced reliance on oil and gas production and exports, and economic growth may continue even in the face of substantially reduced hydrocarbon income.

The need for oil exporting countries to diversify their respective economies and progressively reduce dependence on oil as a source of income has long been recognized and paid at least lip service. Whether justified by the prospect of eventual exhaustion of underground resources – however remote in time this might have been judged to be – or by the fear that sooner or later technological development might make oil less economically relevant (the Stone Age did not end for lack of stones[1]), the need to reduce dependence on oil income has been a mantra in all official and scholarly literature.

Such unanimity has largely hidden the fact that we do not very well know what successful economic diversification should consist of. Besides the well understood difference between growth and development - suggesting that the latter is better measured in terms of human development outcomes such as education, health care, life expectancy, respect for human rights and more – it is not at all clear which economic structure will best be suited to guarantee a future to the oil economies: is it large scale capital intensive industry (be it oil or non-oil related)? Or small and medium scale manufacturing (dependent on expatriate labor)? Or logistical, financial and other services? We witness divergent models and outcomes, with individual countries faring very well or very badly, depending on the exact cocktail of indicators chosen.

The new phase of lower oil prices - which may well be rather long-lived - will provide us with some very interesting insights. As long as terms of trade are favorable to oil producers, the extent and viability of diversification remain clouded by multiple sources of ambiguity. National account indicators have much less precise meaning that is frequently attributed to them, and statistical effects hide underlying realities. High oil prices may exaggerate the importance of oil revenue as a share of GDP, and an expanding bureaucracy may give the impression that non-oil value added is growing simply because government services are accounted for at cost. Once oil prices cool down, especially if in a precipitous way as in 2014-15, a completely different picture may emerge.

Other parameters will become more important. The large deficit that inevitably opens in government accounts will be clearly unsustainable in the long run, even if accumulated reserves may allow reducing it gradually over many years. Excessive energy intensity and almost exclusive dependence on domestic hydrocarbon production will underline the danger of the exportable surplus narrowing down. Acute dependence on expatriate labor will appear increasingly questionable, and unemployment of nationals intolerable.

Today, sustainability is a challenge for all countries, not just for the Arab oil exporters of the Gulf. Climate change, looming water scarcity, or urban sprawl, which leads to disconnect between man and nature, are universal issues. We have technological solutions for all, but social and political realities are not always up to the task. Industrial democracies as well as emerging countries are under stress, and no country can claim to have solved all problems, or expect to isolate itself from the consequences of others' mistakes. In a very real sense, we all need the Arab Gulf countries to succeed.

Annika Kropf steers away from easy generalizations and asks more questions than she provides answers to. This is refreshing and intellectually honest. She concludes that regulatory capacity is essential for transforming resources into economic growth – and this much is not a surprise. She also concludes that democracy is not an advantage for economic growth – at least not in the cultural and political environment of the Arab Gulf. This might not be the conclusion that the reader would like to see, but again is not surprising in the eyes of scholars of the region. Yet authoritarianism may be dangerous and even catastrophic, as many examples demonstrate.

The worst failures in the region are attributable to the irresponsible behavior of authoritarian leaders such as Saddam Hussein or Muammar Gadhafi. Their fall has however not led to greater stability - the contrary is the case. The wind of revolt that swept through the region at the beginning of this decade has led to greater instability and more conflicts; no country can be said to be immune.

How the region will emerge from this difficult passage is not clear. If in fifteen or twenty years the region will have recovered some peace and stability, it is possible that we shall be celebrating the progress of economic diversification, prompted - or forced - by lower oil prices and ever more threatening global challenges. We shall then rightly conclude that the truly necessary precondition for growth and development is just that: Peace.

Giacomo Luciani
Geneva, October 2015

Acknowledgements

At its beginning, this book was a purely academic project that resulted from the fact that, while methodology plays a huge role in the field of political economy, methodology and empirical studies are a neglected topic in Middle Eastern Studies. In the course of writing, I had many opportunities to study the phenomena described in the book qualitatively and on the ground. While I had been aware of the huge impact of the poor data on quantitative studies of the MENA region, work experience in and with companies and institutions from the GCC states taught me to what extent economic success or failure are shaped by behaviors and processes that are neither easily measured nor easily described.

For these experiences, I am indebted to many people who opened doors for me and helped arrange interviews, factory visits, visas and journeys. Faisal al-Muammar, in his function as advisor to the late Saudi King Abdullah, trusted me with several important projects that gave me the opportunity to see places and meet people only very few people have access to. Without Brij Bali, I would not have had access to so many educational institutions and companies in the UAE. Without Tanya Cariina Newbury-Smith, Saudi Arabia would probably still be a white spot on my map, and I am also heavily indebted to her for the thorough proof-reading of the manuscript.

I am also very grateful for all the formal and informal conversations that often taught me much more than books. In this regard, the Gulf Research Meeting in Cambridge, where I presented my research on several occasions, proved to be an invaluable platform. My supervisor, Prof. Hans-Georg Heinrich, opened a new area of research to me, introducing me to recent developments in methodology.

Lisa Pelling helped me a lot in what was an often painstaking process of research design and writing. Many a good idea stems from our joint readings and commenting of chapters of the book in its initial stage. In this regard, my publisher, Kai-Henning Gerlach, and also Giacomo Luciani helped me turn an academic exercise into a more concise and focused book.

Eventually, this book would not have been possible without the help and understanding of my family, especially my husband and my sons, who often had to let me travel abroad or work long hours in the evening.

Note

Transcriptions

Arabic words or geographical names where spelled are as in the Oxford Dictionary. To improve reading fluency, proper names were spelled as they usually appear in the English language press and publications. If these names, especially geographical names, are less familiar or may be confusing, the detailed transcription will be given in parentheses when the word is first mentioned.

If Arabic words could not be found in the Oxford Dictionary, I used the transcription used by the US library of Congress. Foreign language words other than proper nouns or geographical names were italicized. (Download: http://archimedes.fas.harvard.edu/mdh/lcromanization.pdf [last access 10 April 2015])

A glossary of Arabic terms can be found at the end of the book.

Abbreviations

The appendix (Appendix A) contains a list of abbreviations used in this study.

1

Introduction: The States of the Gulf Cooperation Council – The Paradox of Wealthy Less-Developed Economies

For several decades now, the states of the Cooperation Council for the Arab States[1] (GCC) have defied many preconceived ideas of modernity and development. Half a century ago, these countries could hardly be called anything but undeveloped. Illiteracy, subsistence economies and no electricity in large parts of the countries were no exceptions. Oman, to give just one example, was almost cut off from the rest of the world until 1970.

Today, all six states of the GCC are among the first sixty countries of the Human Development Index. (UNDP 2011) Even though they have remained developing countries, they have made tremendous progress.

The term developing country usually evokes a picture of states where a few elites have benefitted from growth while the rest of the society bears the burden of industrialization: exploitative and unhealthy working conditions and low wages, while infrastructure and social welfare are only slowly improving, if at all. The situation in the states of the Gulf Cooperation Council is different. The wealth seems to have reached all layers of local societies. There are not just a few wealthy oligarchs; all citizens seem to have expensive cars, the latest cell phones, access to free health care and other welfare while paying virtually no taxes.

To many Western observers it is puzzling to find these well-known features of our own technological era combined with what could be perceived as traditional and Islamic features: a large percentage of local women can only be seen veiled and the idea persists that women can only leave the house or country with their husband, father, or brother. In Saudi Arabia, for instance, women are not allowed to drive a car and many public spheres are divided into female and male zones. International media keep repeating these phenomena as major signs of discrimination against women, although these rules are often not effective. Equally complicating matters might be the planning of a shopping trip or visits to the authorities according to the daily prayer times during which clients in Saudi Arabia often have to leave a shop, mall or offices.

What might be more puzzling than the seemingly traditional habits of the locals, however, is the fact that one hardly meets and communicates with them. Knowledge of Urdu or Hindi is often of greater value in daily life than Arabic as most services are provided by the huge amount of expatriates in the region. Blue collar work is done by workers from poorer countries such as Pakistan, India, or other countries in South-East Asia, who often send their money back to their families. Due to labor migration, development in the GCC states has an important international outreach that has shaped the economies of labor exporting countries too. In the smaller Gulf States, these workers are more numerous than the local population and "the Gulf economy and lifestyle, as it is now, would simply not be sustainable without its migrants." (Osella 2011: 8) Nevertheless, these workers have to struggle for the recognition of the essential work they do and their working conditions are often precarious. Returning empty-handed to their home countries, however, is not an option either.

The "locals" prefer to work in white collar jobs, but also here, many high-skilled posts are held by foreigners from industrialized countries or other Arab countries: the latter work as engineers, teachers, doctors, professors or advisers. Gulf citizens – and this is not only a prejudice – mostly work as civil servants in the inflated public administrations. (Davidson 2009: 149-153)

Although these impressions from the region are not unjustified and are supported by a vast amount of data, they are not always 100% clear. Modern vs. traditional, male vs. female, foreign vs. expatriate are helpful, but often deceptive categorizations.

Many of the traditional features, especially where women are concerned, are the result of a multi-layered interplay between traditional and Islamic institutions and the rapid modernization in the aftermath of the oil windfalls. They are far from being mere remnants of old traditions, and some of them, such as women who stay at home instead of work for pay, have only been made possible by the revenue from oil. Women are now better educated than their male peers in the GCC, and staying at home or wearing the *abāya, hijāb* and *niqāb* is sometimes a very modern choice, and not a traditional or Islamic custom. The same holds true for business habits which have become a mixture of foreign influences and traditional habits. Businessmen may wear the traditional *dishdasha* or *thobe* and conversations and negotiation styles may differ substantially from Western habits (Cole 1992; Badran 1995; Hecht-El Minshawi 2007; Rothlauf 2006; Berger et al. 2014) but nonetheless, Western influences, often imported via consultants and expatriates, are strong and often cause confusion, not only between foreigners and locals, but also between the younger and older generations, or local and foreign-educated nationals. It is not unusual that young Gulf nationals become demotivated in inefficient, highly hierarchical business structures, and one has to ask whether the oft-quoted rentier mentality has become more structural than mental.

The same is true for other ideas about the "local". While it is true that one would hardly find a Gulf citizen working at a building site, there is poverty, unemployment and underemployment among nationals of the more populous GCC states. Even employed, well-educated young men often find that their skills cannot develop within the local business structures, whereas others face an unbridgeable gap between fulfilling the traditional role of a son in an extended Gulf family with the demands of a modern labor market, such as spending time working or studying abroad. Young males especially have been the subject of a lot of concern in the fast-growing Gulf societies.

Nonetheless, the puzzle of seeing wealth and industrialization grafted upon traditional societies remains. In the industrialized countries of the West the relative wealth today is the result of a long development process whereas in the GCC states it stems mostly from oil exports. There is nothing in the GCC states that has not been affected by the discovery and production of oil and gas. Natural resources are not only the origin of all wealth and luxury, but several decades ago they played a major role in the consolidation of the young and fragile states and the drawing of their borders. Today, these resources are the basis of power of the ruling dynasties and therefore crucial for the maintenance of the *status quo*.

However, Gulf States may sooner or later run out of oil and gas, as they are depletable resources. When and if this occurs is unclear. New oil and gas wells keep being discovered and production techniques continue to be better developed. There are experts of the oil sector who forecast an economic rather than "real" depletion: at a certain point, oil might be replaced by other sources of energy because its production (and its negative externalities) might become too expensive. Others warn that peak oil has already been reached or will be reached soon. (Heinberg 2003; Deffeyes 2010; Simmons 2005; Legett 2006; Legett 2013) While oil prices may rise tremendously before final depletion, most scenarios point to the possibility that oil and gas exporters will eventually have to adapt to less income from their natural resources – be it due to lower prices or depletion. Recently, the massive expansion of shale gas extraction, mainly in the US but also in Europe, has caused immense uncertainty as to how demand for fossil fuels would develop. However, even in absence of such a scenario, the dependence on natural resource exports has downsides as it makes these states vulnerable to exterior influences that become manifest in oil price fluctuations. The decline of oil prices in the 1980s resulted in serious difficulties as well as budget deficits for oil exporters such as Saudi Arabia or Kuwait and this scenario seems to repeat itself in 2015 with an oil price well below US $60. After the oil price boom in the 1970s, tremendous sums poured into the construction sector and the welfare system. Free education and health care and jobs in the public sector have diverted people from participating in politics. Gulf citizens have come to expect their share of oil windfalls, making it harder for governments to cut back on spending. (Beblawi 1987: 86) The latter have therefore tried to prevent a decrease of revenues by following two main strategies: the first consists in rent diversification and maintenance. This strategy includes policies to guarantee a reasonably high oil price, with an accumulation of alternative sources of rentier income, such as foreign assets.

The second strategy is what one may call "non-oil growth" or "non-depletable growth"[2], corresponding to the growth of the non-resource sectors of their economies. This is part of the diversification strategy that mostly consists of downstream projects, such as processing oil and gas further, although upstream projects could also be part of such a strategy. The question of whether this can be considered "diversification away from oil" is disputed. Most of these projects still rely on oil, gas and other natural resources, but so do most industries all over the world. Leaving the natural resources in the soil is not a real alterative either. As Luciani put it (2012: 184), "the sensible approach to economic diversification is not *away* from oil but *leveraging the availability* of oil and gas."

Rent diversification and rent maintenance have been quite successful so far: the revenue from capital investments abroad has provided substantial income. The sovereign wealth funds of the GCC states have so far operated very successfully, partly due to the support from the world's best investment experts. They also have a reputation of being interested in long-term investments rather than in short-term speculation. Recently, the financial crisis (2008) has probably decreased these assets but also opened up more possibilities for international investments. (El-Kharouf; Al-Qudsi; Obeid 2010)

As for rent maintenance, where OPEC succeeded in keeping the oil price high in the 1970s, prices plummeted in the 1980s. OPEC's production agreements proved rather ineffective in the aftermath. (Ramady 2015) The recent oil price rises suggest that factors other than OPEC have far more impact on oil prices. Today, oil prices seem to be largely out of the producer's control and more heavily influenced by international political events and financial speculation.

Diversifying economies away from the oil and gas sectors, however, has turned out to be very difficult despite huge investment capacities. And even if some countries have made more progress than others, the economies of the GCC states still rely heavily on oil revenues for subsidies. The quest for "non-depletable growth", therefore, has to continue.

While rent maintenance and diversification may be as important and as valid as non-oil growth, this study will only deal with the latter. The following three questions will be asked:

1. Have the GCC states made significant progress in developing their non-oil economies?
2. What factors and which combination of factors account for the success or failure of non-depletable growth in the GCC states?
3. What role does the degree of resource abundance or dependence play in this process?

The first question may be more difficult to answer than expected. Official data and growth rates more than ever hide the real economic development of oil exporters. (Beutel 2012) It is necessary to go beyond ordinary macroeconomic statistics in order to judge whether the GCC states are on a path toward becoming less resource dependent economies or not. In the end, success depends on more than the revealed data. It is impossible to predict whether the entire process of diversification is sound and would shield the GCC states from trouble were income from oil and gas to run out. There are cases of rich natural resource exporters such as Nauru that have seen their sovereign wealth funds depreciated and their resources depleted, leaving them with just spoilt administrations and environmental problems. As Piketty (2014: 7) wrote in his famous *Capital in the Twenty-First Century*, it "is much too soon to warn readers that by 2050 they may be paying rent to the emir of Qatar". This study, therefore, will only focus on one aspect of this strategy. It is important to note that this question is focused on the very aspect of the non-oil economy of and *in* the Gulf states. I do not analyze the overall development process or diversification, partly because I doubt that it is possible. Several projects depend on international developments which are hard to forecast, and several development strategies — such as debt-financing in Dubai — are heavily debated, and judging them as good or bad is an ideological question. Therefore I will judge the economic performance by the local development of the non-oil economy. This choice was necessary because reliable information on foreign assets is hard to obtain. While some funds operate transparently, others are rather opaque. Sovereign wealth funds (SWFs) in some GCC countries are only a comparatively recent phenomenon. Often they are meant to be a fund for future generations but it is unclear if and how such redistribution would work.

The answer to my first question is also essential to my second question: what factors and what combinations of factors account for the success or failure of non-depletable growth in the GCC states? Why do some countries perform better than others? As I will show in my theoretical overview, these are very common questions in growth theory and development studies, but it seems that large-scale oil exporters are exceptions.

As a third consequential question, this study also adds to the resource curse debate. Do very resource-rich countries really grow more slowly than less resource-rich countries? And if so, is there a threshold between resource-poor and resource-rich countries that marks the beginning of a resource curse? This question will be answered with regard to the six countries in question. Economists have frequently investigated economic growth in developing countries and no single remedy for bad performance has been found so far. Lessons learned in industrialized countries have often been applied to less developed countries, and failed. (Onyemelukwe 2005: xi-xv; Easterly 2001). Instead of applying growth theories of industrialized countries to the GCC states, I would like to investigate the specific patterns of growth in this very homogeneous region. Such a focus also allows and even calls for a more thorough discussion of their economic, social, cultural, and political conditions.

The various conditions, in turn, raise the need for a sound theoretical background of the variables. The choice of a theoretical framework has to take into account one important paradox: the GCC states are wealthy countries, but nevertheless their economies are underdeveloped.[3] This makes it difficult to decide whether we must deal with economic *growth* or *development*. Usually, studies of economic growth deal with growth in developed countries whereas development economists try to find out how less developed economies *could* grow and become developed. Moreover, development should also contain the aspect of a qualitative improvement. Herman N. Daly, for instance, contrasts growth and development as follows:

> 'To grow' means to increase the size by the accretion or assimilation of material. 'Growth' therefore means a quantitative increase in the scale of the physical dimensions of the economy. 'To develop' means to expand or realize the potentialities of; to bring gradually to a fuller, greater or better state. 'Development' therefore means the qualitative improvement in the structure, design and composition of the physical stocks of wealth that results from greater knowledge, both of technique and of purpose. (Daly 1990: 113)

This qualitative aspect is also interesting to the Gulf States, not because growth in general is needed, but non-oil growth is required.[4] Thus, development theories seem to be more likely to explain this process. However, development theories suffer from a huge disadvantage compared to growth theory: in developed countries growth has already taken place and can be analyzed in hindsight. On the other hand, where development and therefore stable and significant growth has yet to take place, advice on development is rather difficult.[5] Such advice could only come from other countries' experiences, and these are likely to have a different outset. What about the GCC states in this context? In some, per capita GDPs are higher than in the USA. If their growth rates are computed with GDPs that include oil exports, one cannot deny that *growth* has taken place. In many instances therefore, the GCC states do not share the lot of many underdeveloped countries.

The case selection puts this study between two major and overlapping theoretical environments. To further confuse, both have shaped development policy. Models of development theory, and growth models that have been developed in a European or American context have also been applied in development policy. Meanwhile, growth theory has given rise to a huge amount of empirical literature, mostly using regression analysis. (Barro 1997; Mauro 1995; Quinn/Woolley 1995; Ram 1986; Rodriguez/Rodrik 2000; Sachs/Warner 1997; Romer 1994 and many others) Unfortunately, this kind of research on economic growth has not resulted in any reliable inferences. While some variables seem to be positively correlated with growth in a first analysis, the next paper often claims the opposite. (Quinn/Woolley 1995; Stiglitz 2005)[6]

Development theory, on the other hand, draws attention to more than just economic factors and takes into account that development today cannot follow the same patterns as development one hundred and fifty years ago. Therefore, development theory challenges assumptions such as perfect competition within a country and a closed economy. As for perfect competition, few economists would deny that even markets in developed countries have certain imperfections. This is all the more true in LDCs where institutions, which should guarantee competition, are likely to be weak.

This is one of the critiques claimed by "New Institutional Economists" (New Institutional Economics, NIE) who state that neoclassical growth theory took many conditions for granted that nevertheless are far from being fulfilled in less developed countries. They emphasize the role of institutions[7] in guaranteeing property rights, contracts, and in general, the compliance with rules. Thus the NIE made a few steps into the direction of the social sciences, which are often considered too "soft" by economists. (Clague 1997; Harriss 1997)

As this study will reveal, neither growth nor development theory can fully explain non-depletable growth in the GCC states. This failure of universal theories to explain specific cases calls for taking a closer look at the sample countries under scrutiny in this study. Their most important features are the vast oil reserves, Islam, and some characteristics of Arab societies.

Economies based on natural resources have been the subject of several theories and studies. A rather economic, but geographically focused approach to the topic of resource abundance and growth, is given by Richard M. Auty's work on *Resource Based Industrialization* (1990; see also Auty 1993; 2001; 2006). Auty's work is strongly related to two disputed phenomena that are always present in literature on resource abundant states: the resource curse (or "paradox of plenty") and Dutch disease. (Looney 1991; Basedau/Lacher 2006; Esanov/Raiser/Buiter 2001; Gelb 1988; Karl 1997; Palley 2003; Sachs/Humphreys/Stiglitz 2007)

The resource curse is a phenomenon rather than a theory or an explanation. It describes the experience that resource-rich countries tend to grow slower than their resource-poor peers. The terms "curse" and "paradox" suggest that economists find it hard to explain this phenomenon.[8]

Sachs and Warner sum up the attempts as follows: "Most current explanations for the curse have a crowding-out logic. Natural Resources crowd-out activity *x*. Activity *x* drives growth. Therefore, Natural Resources harm growth." (Sachs/Warner 2001: 833)

For Sachs and Warner, one explanation is the Dutch disease. In a nutshell, Dutch disease claims that the large availability of natural resources and their exportation lead to a decline in the manufacturing sector, which becomes less attractive due to higher rents in other sectors. (Gylfason 2001a)

Political scientists focusing on states that receive huge revenues from natural resources have coined the notion of "rentier states". The rentier state model establishes links between rents (often defined as income generated without any effort of society), political authoritarianism, neopatrimonialism and economic inefficiency.

Pioneering work on rentier state theory was published by Giacomo Luciani and Hazem Beblawi in 1987: *The Rentier State* is a collection of case studies and theoretical works exploring the interplay of rents, the state, and the society. Later on, the theory was supported and amplified by Peter Pawelka, who looked for an historical explanation. In his view, the influence of colonialism on the local economies in the Middle East explains part of the rentier state theory because local manufacturing was partly destroyed while resource exports were enhanced. (Pawelka 1993) Recently, Martin Beck (2007) looked into cases which allegedly proved the rentier state theory to be wrong and specified the theory further.

In addition to these theories, a lot has been written on culture and economic growth as well as on Islam and economics. While recent literature on "Islamic economics" is usually either descriptive or propaganda for the implementation of certain Islamic economic institutions, opinions still differ widely on the relationship between Islam and growth. (Aggarwal/Yousef 2000)[9] The classics on these topics, such as Max Weber's *Die protestantische Ethik und der Geist des Kapitalismus* (1905/2006) or Maxime Rodinson *Islam et Capitalisme* (1966), contrast with both Orientalist writers on the Middle East — who claim that Islam is opposed to modernization — and Muslim theologians — who often hold that truly lived Islam would automatically lead to economic well-being. In this regard, I will follow an institutionalist approach developed by Timur Kuran, who has investigated the relationship between Islam and underdevelopment for several decades now. (Kuran 1997; id. 2008; Tibi 1985; id. 1991)

Eventually, while Islam has shaped many institutions in Arab societies, some features predated the advent of Islam. One could even argue that Islam failed to supplant them, as in the case of kinship in the Arab Gulf. (Alsharekh 2007; Gellner 1992) These kinship structures are characteristics of a collectivist society which has been contrasted to an individualist society regarding its consequences for economic behavior by Avner Greif. (Greif 1994)

All these theories will have to be taken into account to find the conditions and constellations of conditions that impede or accelerate economic growth in oil exporting states, with a special focus on the GCC states.

At the beginning, a reasonable measurement of the outcome/dependent variable is necessary: the growth of the non-oil sector of the economy, which is one part of economic diversification. Even if non-oil growth contains a normative aspect (it is "necessary") and is a complex phenomenon if it really aims at making the economy more independent from the oil sector, a simple and feasible operationalization has to be found.[10] To calculate these growth rates, a non-oil GDP has to be calculated in order to filter out the non-depletable growth rate from the real growth rate to be found in the statistics that is – in times of rising oil prices – much higher. Thomas Stauffer and Lennox have presented a sophisticated method for measuring income in Arab states. (Stauffer 1987; Stauffer/Lennox 1984) However, data quality makes following this method unrealistic. I have thus decided to use the World Bank method: subtracting natural resource rents from GDP and calculating new growth rates on this basis. (Neumayer 2004: 1630) This is a huge tradeoff but it represents the possibilities at the beginning of this work[11] as well as the necessity to find comparable measures, not only for the GCC states but also for all other countries in the regression sample.

The question of what factors and which combination of factors account for the success or failure to develop a non-oil economy seems to be a standard case for regression analysis. I will also use a regression that did not produce sound explanations. It is complemented by two more approaches that will form a methodological triangulation (Flick 2004) on three levels: The case level (case studies of the GCC states), the small-n level and the large n-level (regression).

Figure 1 **Methodological Approach**

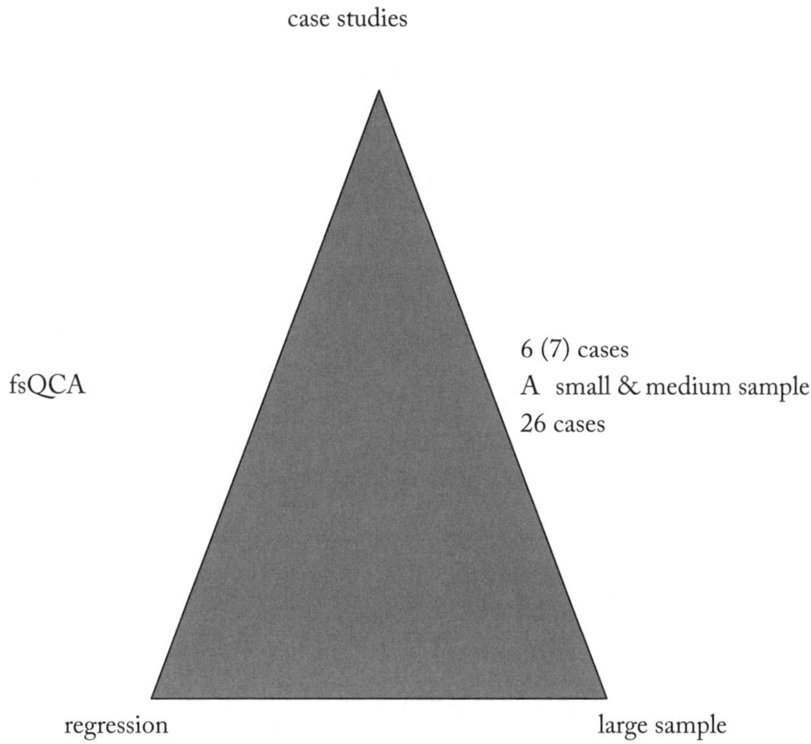

While the case level and the large n-level follow common methodological practices, the intermediate level calls for further explanation.

Many shortcomings of regression analyses can be avoided by choosing a tool that comprises qualitative and quantitative methods. Charles Ragin[12] developed an analytical tool that allows for medium-n samples (5-50 cases) and subset analysis: Qualitative Comparative Analysis (QCA). Instead of probabilistic theories, it works with Boolean algebra which is known for its binary logic: conditions can be true or false. However, one version of QCA, Fuzzy Set QCA (fsQCA), does not use these crisp sets (yes or no/true or false decisions, as in a truth table), but allows variables to adopt values between 0.0 and 1.0. (Ragin 2000)[13] Instead of relying solely on data, software and correlation coefficients, Ragin's method is rather a tool to see cases more clearly and to compare more than two cases on a deeper level. The researcher's qualitative knowledge about the theories and single cases still plays a very important role.

The sample size of six countries is at the lowest limit of a QCA analysis. And it should be clear that with such a small sample one can only contribute very little to modification or building theory. Therefore, this study relies on two QCA analyses: one comprising the GCC states and another one embedding the GCC into a larger sample of twenty-five oil-rich or resource rich countries.

The operationalization of the variables will rely on data as well as on qualitative assessments. For economic conditions such as the investment share of the GDP, oil price or oil reserves, I will use widely used databases. Many important scientific data sets are available on the internet such as the PENN World table (Summers & Heston data set), the data sets by Warner & Sachs, Barro & Lee or Michael Herb.[14] The International Monetary Fund (IMF), the World Bank, the Organization of the Petroleum Exporting Countries (OPEC), Transparency International as well as the various organizations of the United Nations, also provide useful data.

Some factors of theoretical importance, however, cannot easily be measured. For some cases, listing several relevant factors and looking for their absence or presence is used as a substitute. Several variables/conditions were compared to qualitative findings, even if they were already based on data. However, attempts to use qualitative expert interviews failed to produce reliable results. An example is my attempt to gather data on "rentier mentality" or any aspect of working mentality. Fourteen expatriates who had worked in the GCC countries as well as six experts on the region have been interviewed online to assess the degree of rentier mentality in their host countries. In too many cases, two experts of the same country had contradicting views on the prevailing work mentality. It seemed that experts or expatriate workers with a rather short-term experience tended to opt for harsh judgments on a countries working mentality, while expatriates with long term experience seem to have gained a deeper understanding, leading to the refusal to judge the country in question by Western business standards. Talking to Gulf nationals, issues like rentier mentality caused irritation and resentment. This is why some important aspects, like work-mentality, will not enter the analysis as a variable.[15]

Nevertheless, expert interviews were conducted during a field study in Abu Dhabi and Dubai in February 2009 and a stay in Saudi Arabia in 2014. Conferences on and in the Gulf States have given me further opportunities to discuss this study with researchers from the region.

The methodology leads to a hetereogeneity of research styles that will also be reflected in the structure of this study.

The following two chapters deal with the regional background and the theoretical basis of this study. Chapter two gives a short introduction of the main characteristics of the region and characteristics of oil-rich countries, followed by the case studies. How the situation in the GCC states challenges common growth and development theories is the object of chapter three. In addition, theoretical implications of some of the region's characteristics will be discussed at the end of the chapter. These include theoretical considerations of Islam, economic growth, and women's issues. Chapter two and three will conclude by assessing the relevance of the theories explained in chapter three for oil exporting states, and the choice of the variables.

Chapter four will provide an introduction to fsQCA and its application to this study. The same chapter will also explain how the variables will be operationalized and calibrated. The analyses will be conducted in chapter five, starting with a standard cross-country regression. This analysis will be viewed from the GCC angle. In a next step, the GCC will be picked out as a sample of its own and analyzed using fsQCA. Necessary and sufficient conditions of positive and negative outcomes (non-depletable growth performance) will be identified, followed by a discussion of their consistency and coverage. The "best" variable constellations will be discussed and the implications of the method will also be highlighted. As a final step, the GCC states will be analyzed in the larger sample of resource-rich countries.

The discussion in chapter six reviews the results of chapter five in light of the theories, while the conclusion highlights implications for the current and future economic performance in the GCC states.

This study was mainly conducted and written between 2008 and 2011. The data used in the empirical analysis is the data that was available at that time. Given the complexity of the models and the different methods used, using more recent data would have meant changing not only the data, but also the entire calibration. At the same time, 2007/2008 marked the beginning of the global financial crisis that added a new, international dimension to the growth performance of the GCC states, and of the so-called "Arab Spring" events. Although these topics will not be omitted from this analysis, it is important to note that the data used in this study is gathered from before the global financial crisis and the Arab spring had left their mark on the region. There are other studies covering a shorter time frame that cover this topic (Ramady 2013).

What is more, this study should be perceived as a first step towards a more systematic and methodologically-focused approach towards the political economy of the Middle East, which is still mostly a domain of qualitative research. Given the huge data problems and, apart from a few exceptions the lack of viable economic theories tailored to that region, the time may not yet be ripe for such an approach. However, this study may nevertheless point to the improvements required to allow for this kind of research in the future and it is hoped that these improvements will step-by-step allow for more suitable operationalizations and models.

2

The States of the Gulf Cooperation Council: Historic, Economic, and Political Background

The Cooperation Council for the Arab States of the Gulf was established in 1981 and its six member states are Bahrain, Kuwait, Oman, Qatar, Saudi Arabia and the United Arab Emirates. These countries have a lot in common. Their inhabitants are mostly of Arab origin, they are traditional monarchies and they are marked by a conservative official interpretation Islam. None of them has an outright colonial past that could be compared to Algeria or India. With few exceptions, there was little interference in internal matters and partly as a result their inclusion into the world economy took place very late. (Pawelka 1993)

Nevertheless, there are differences of a "quantitative" nature. Their oil and gas reserves vary significantly and so does the size of the countries. Therefore, this chapter will provide background information on the GCC states, followed by a description of the Gulf Cooperation Council itself. Given the fact that all six states are important oil-exporters, a chapter on the characteristics of oil based economies will precede the case studies.

This book does not contain any historical abstracts but there is a growing literature that fills this gap. A work of reference has been provided by Rosemary Said Zahlan's (1998) *The Making of the Modern Gulf States: Kuwait, Bahrain, Qatar, the United Arab Emirates and Oman*; also Jill Crystal's (1995) *Oil and Politics in the Gulf. Rulers and Merchants in Kuwait and Qatar*. On Dubai and Abu Dhabi, Christopher Davidson (2008; 2009) has recently published separate and detailed volumes. On Qatar, Allen J. Fromherz, *Qatar. A Modern History*, was released in 2012. As for Saudi Arabia, Madawi al-Rasheed's *A history of Saudi Arabia* (Al-Rasheed 2002) is a good reference. (Al-Rasheed 2002) Francis Owtram (2004) provided a history of Oman: *A Modern History of Oman. Formation of the State since 1920.* These volumes take also into account economic aspects.

A few decades ago, research on the GCC economies was mostly published by a small group of authors including Giacomo Luciani, Hazem Beblawi, Rodney Wilson, Tim Niblock, and Robert Looney (1994; 2004). During their long publishing career they wrote and edited numerous volumes on aspects of political economy. Recently, a new generation of prolific authors has come into being. Examples are the recent volume edited by Jean-François Seznec and Mimi Kirk (2010) and especially the numerous publications by Steffen Hertog. Another ambitious and promising contribution is the four volume paper collection by Luciani, Hertog, Woertz and Youngs (2012) which is still in the course of being published. The Kuwait Programme on Development, Governance and Globalisation in the Gulf States (mostly at the London School of Economics) has given numerous young researchers the possibility to publish mostly on recent developments, diversification issues included (Shochat 2008; al-Kuwari 2013; Hvidt 2013; Tenreyro/Koren 2010) Apart from these sources, I have used journals, newspapers, assessments by the IMF and the World Bank as well as information from investment guides such as the Oxford Business Group and Germany Trade and Invest. (2011) Embassies often provide information for investors on their websites. For information on the situation of the 1980s and 1990s, I had to rely on qualitative literature or investment guides. (Herzog 1998)

Some of the GCC states are well covered by the literature. This is especially the case for Saudi Arabia, the emirates Abu Dhabi and Dubai, and to a certain degree Kuwait also. Fewer publications are available on Oman, Qatar, and also Bahrain, although their numbers are rising too.

2.1 Commonalities: Islamic "Oil Monarchies"

A big part of the world's proved oil and gas reserves are located in the "oil monarchies" (Gause 1994) of the Persian Gulf. Estimates of the R/P (reserves to production) ratios are much higher for the Middle East than for other regions of the World (British Petroleum 2013), indicating that oil would only be depleted within several generations if production rates remained stable.

The oil market is influenced by a whole array of (geo)political, social and economic aspects, the psychology of the market playing a huge role. Also, supply is the result of a complexity of aspects and processes, making it almost impossible to forecast. Revenue and profit maximizing of oil revenues is a difficult task, as is forecasting developments: the oil market can neither be called free and (perfectly) competitive, nor monopolistic. In the first half of the 20th century, the market was controlled by a few big companies (known as the "Seven Sisters"). When oil-exporting countries later on revoked the oil concessions and nationalized their oil industries, oil producers increasingly lost control over the market. OPEC managed to reach higher prices only for a short time, until the market entry of non-OPEC oil producers, as well as demand elasticity in oil consumption, made the market

Figure 2 **Oil R/P Ratios by Region as of 2006 and 2012**

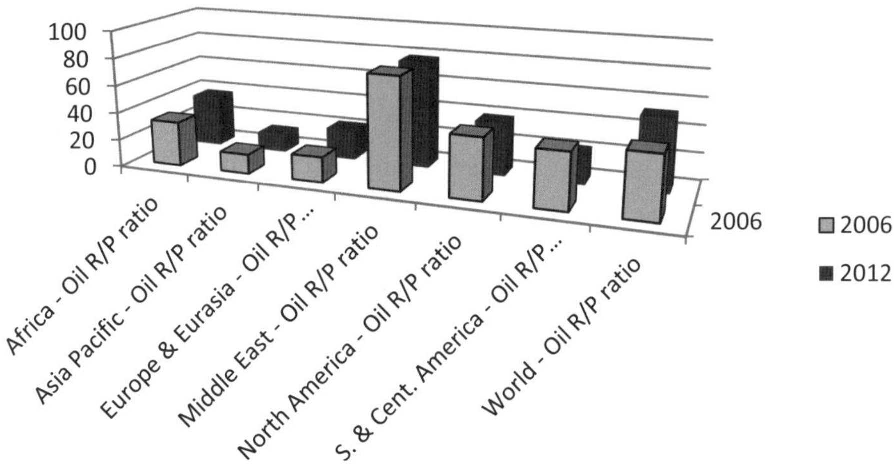

Source: BP 2013; data for Venezuela was missing in the 2012 calculation for South and Central America.

unpredictable. Furthermore, especially in OPEC states, oil is regarded as a subject of national interest, and as the oil wealth is exclusively managed by governments, they have more than purely economic motives for decisions such as political legitimacy or foreign policy. In other cases, attempts to increase market shares or to reach desired prices backfired. This has often been the result of Saudi Arabia's OPEC policy. (Amuzegar 2001: 10-20 and 40-46)

It is also noteworthy that oil exporters do not necessarily have shared interests. While an oil exporter with rapidly decreasing reserves might prefer extremely high prices because it would like to gain a maximum from the little oil left, large oil exporters with ample reserves have no interest in very high oil prices that encourage the development of substitutes.

The use of natural gas has dramatically increased over the last decades. Unlike oil, gas cannot be easily transported. In oil exporting states, this means that the local consumption of electricity and hydrocarbons should be met by gas rather than by oil. For transport, the construction of pipelines is necessary. If transported by ship, gas is usually liquefied by either cooling or by compression. This requires a lot of capital investments and infrastructure. The last decades have seen the increasing use of gas (British Petroleum 2007), the construction of pipelines, and of special ships and ports for the shipping of natural gas.

It is noteworthy that the endowment of oil and gas contrasts sharply with a lack of arable land and fresh water in the GCC states. (FAOSTAT 2011) The scope for agriculture is extremely limited and costly, since water has to be desalinated in energy-costly desalination plants.

All six GCC economies still rely on oil and gas. Short-term price volatility and long-term uncertainty as well as possible future depletion make this fact problematic. In addition, the oil sector is capital intensive but not labor intensive. It can barely absorb the young,

Figure 3 **Development of Natural Gas Consumption (1965-2012)**

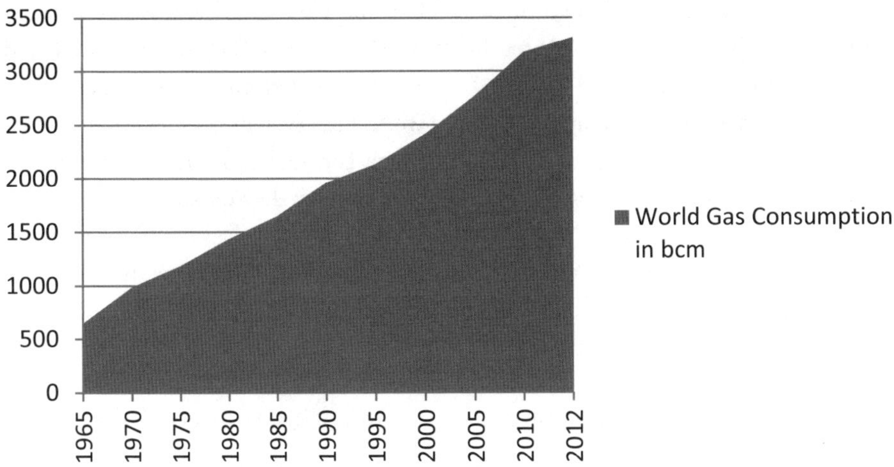

Source: BP 2013.

rapidly growing labor forces and demands a small number of highly skilled experts, and the necessary blue-collar work force must not be averse to hard and sometimes dirty work. The linkage potential of these sectors is very limited, and in most cases, the "fiscal" linkage – government investing the income from oil in other industries – is the most important one. (Amuzegar 2001: 17)

There are two main ways to confront these problems. The development of the local economies, and the investment in foreign assets. The disadvantage of investing into foreign assets is evident: save for some employment opportunities in investment, no local work opportunities will arise. Furthermore, this avoids rapid social change and perhaps unnecessary environmental pollution. Limited absorptive capacities might also lead to a surplus even after investing into local development, making investment into foreign assets quasi inevitable. What is more, money invested abroad is not lost to the local economy. Depending on the investment and the liquidity of the investment, it can be used for investments whenever they appear necessary.

There is therefore a point of not investing too heavily into local development at once. Nevertheless, some GCC states made this choice: infrastructure investments were especially enforced, even if it was not quite clear what kind of infrastructure would be needed. In some cases over-capacities were created. According to a widespread opinion, a viable infrastructure would help to avoid development bottlenecks later on. The massive investments are evocative of "Big Push" theories of huge development planning: infrastructure, administration, primary industries, housing, airports, ports and entire cities were created at the same time, in the hope that this would encourage private sector activities, and development in the secondary and tertiary sector. Investments went to the limits of absorptive capacities. Later on, several projects turned out to be unnecessary or not appropriate, a reminder of Hirschman's critique of the Big Push: investments cannot be made all at the same time because necessities are only revealed by experience. Careful step-by-step financing of local projects as well as the investment of the surplus abroad – instead of hastily initiated projects at home – might lead to the best mixture of making oil revenues work. However, in most cases, the ratio of local and foreign investment is not the result of such considerations. It rather seems that investments were mostly driven by the availability of money: they took place when surplus money was available, and stopped or slowed down when low oil prices decreased the inflow of rents. The importance of the oil price for many economic indicators and investment behavior remains and has been shown in empirical studies. (al-Kuwari 2013: 22-33) The current oil price low (January/February 2015) confirms this observation. Investments are again being postponed.

The choice to invest into the oil-based industry was led by the expectation that the end products would be processed further by the private sector. Furthermore, industries such as petrochemicals, fertilizers, cement or aluminium were chosen, because they are capital and energy intensive. Both are available in the GCC states. (Mueller 2012) For underdeveloped economies this is a new situation, which Nagi described as "development with unlimited supplies of capital". (Nagi 1982) However, this comparative advantage may be mitigated by other factors: Robert Looney points to problems such as the lack of experience, of access to the interesting markets and the absence of both skilled and unskilled labor. (Looney 1990) The strategy of industrialization has to be questioned, and as early as in 1982 Nagi proposed that the GCC countries should instead produce consumption goods in order to substitute some imports. (Nagi 1982) This, however, only happens very slowly and the balances of payment of the GCC are still much less positive than they could be because imports keep rising. As a result of this development planning, which usually takes place in clusters of five years, the economies are marked by huge public sectors. The oil and gas rents accrue to the state, and thus it is the state that has the means to invest in infrastructure and industrial downstream projects. In several cases there are parastatals, closely linked to the ruling family, and the oil industries in particular, of paramount importance to the governments, were nationalized during the 1970s. Many state-owned companies have become successful because they are important enough to receive direct access to the respective ruler who enables them to circumvent slow and inefficient bureaucracies. (Hertog 2012a) Hopes that massive investments by the government would encourage the rise of the private sector, however, have not become a reality so far. The private sector, rather than providing the state with income from taxes that would benefit the entire society, remains government financed and subsidized. Hence, businesses play an entirely different role in the GCC states. They not only benefit directly from the state, demand for their products is also financed by the state purchasing. (Herb 2009)

By employing more nationals, the private sector could play a more structural role in the economy. (Hertog 2013: 51f) However, it is questionable that the current capital-intensive industry is suitable for involving the population actively into the economy. Most heavy industry projects had indeed been government financed because few private investors would have had the means to do this alone. Additionally, Gulf citizens are not eager to become entrepreneurs with their own business ideas and the risk this involves. (Kolb 2015) Exceptions are traditional occupations such as trade and imports. Even fewer Gulf nationals today would work in a blue collar industry. As a result, the governments import labor on a large scale. Furthermore, they employ their citizens in administrative or state enterprises, and it is unlikely that a spirit of entrepreneurship will arise in such situations, although the reasons are more complex. What is more, only limited contributions to diversification can be expected from these industries. If the competitiveness of an industry in the Gulf relies on cheap electricity and capital, this comparative advantage will fade away with dry oil wells.

Given these problems, investments abroad seem to be a viable, even necessary option. Lacking sufficient local investment opportunities, excess liquidity has to be invested elsewhere. Investment decisions can be made by employing the world's best specialists, and given that the GCC states are mainly interested in long-term investments, they are mostly welcome as investors. Nevertheless, the decision of where GCC funds should be invested has to be carefully decided. Investment in outright unIslamic companies (gambling, alcohol, pork) may be not acceptable for the society. Investments in some countries may be frozen in the event of political problems, e.g., in nations related to terrorism. Several funds counteract these risks by operating rather secretly. Nevertheless, there are always currency risks.

Meanwhile, all GCC states have established sovereign wealth funds. The Sovereign Wealth Fund Institute defines these funds as follows:

> A Sovereign Wealth Fund (SWF) is a state-owned investment fund composed of financial assets such as stocks, bonds, real estate, or other financial instruments funded by foreign exchange assets. (Sovereign Wealth Fund Institute 2013)[1]

Usually, these funds operate quite effectively and in several GCC states income from investments abroad exceeds income from oil exports. (El-Kharouf/Qudsi/Obeid 2010)

Another feature all six Gulf countries share is Islam, although there are different sects: Sunnis, Shiites, and Ibāḍis. As elsewhere, these sects are sometimes hostile to or ignorant of each other. However, the bigger rift seems to be between religious and liberal or technocratic forces. The influence of Islamic practices such as Ramadan, prayer time or gender separation differs greatly from country to country and even from region to region. And often, seemingly Islamic features are less Islamic and more cultural than the Western observer might think. The high birth rate in the GCC states is certainly a challenge for the labor market but it has steadily decreased with development (see table 1) and is not higher than in non-Islamic developing countries.

Table 1 Births per Woman and Women in the Labor Force

	Bahrain	Kuwait	Oman	Qatar	Saudi Arabia	UAE	Germany	Botswana	Malaysia	USA
Births per Woman										
1972	5.9	6.9	7.2	6.7	7.3	6.4	1.7	6.8	5.1	2.0
2005	2.3	2.4	3.4	2.9	3.8	2.4	1.4	3.0	2.7	2.1
Percentage of Women in the Labor Force										
1980	10.6	12.7	14.7	8.7	7.9	5.1	39.2	45.4	35.2	41.3
2005	18.8	25.4	16.4	13.8	15.2	13.4	45.2	41.8	35.8	46.2

Source: Word Bank 2011.

What is more, the seemingly Islamic dress code of women does not mean that they are absent from public life. Christopher Davidson (2009: 131f) argues that this "traditional" clothing has become more of a national uniform than a sign of one's religious believe. The fact that today most women wear the black *abaya* and most men the white thobes might be an attempt to distance and to distinguish the national population from the many foreign workers. According to Colton (2011), the gap between nationals and foreigners is one of the most important differences, shaping social stratification in the GCC states. Equally if not more important is the access to the royal family and the government, which, in general, only nationals can reach.

As another commonality, no GCC state can be called democratic even if there are significant differences in the degree of participation. Political scientists usually assume that this lack of representation is compensated for by distributive measures. The state provides its citizens with health care, education, generous unemployment benefits, and financial support for those who earn little money. While welfare systems are also common in Europe, there is a big difference: in Europe, these allocations are funded by taxes, while the GCC governments do not extract these funds from its wealthier citizens. Most GCC citizens today have no experience with an extractive state. They have enjoyed privileges their entire life, and have come to take it for granted that labor migrants, who have to live on much less, do much of the work in their country. The consequences of this redistribution are obvious. While these measures have reduced poverty, they also have reduced incentives. The incentives to work in the private sector are extremely low as long as the government offers well-paid jobs in the civil service. Surveys have shown that work motivation among GCC citizens is consistently lower than among expatriates. (Jones 2015) Although this might be due to a lack of incentives, it should also be noted that highly hierarchical business structures, the lack of open critique, and the lack of a constructive handling of mistakes do not increase work motivation either. (Drummond/Al-Anazi 1997; Hofstede 1980)

The biggest problem, however, is how to go back on these measures once they become too costly. Fast population growth and declining oil revenue might result in budget constraints and it is all too likely that governments will face huge opposition when trying to abolish some subsidies. In Bahrain, this is currently taking place. Bahrain was the first of the GCC states to introduce a very low income tax in spite of fierce opposition from trade unions and religious leaders. (DPA 2007) Decades after these welfare systems have come into being, the results are alarming: many Gulf citizens are unemployed even if they are capable of working. However, they would not accept a blue-collar job and they are not skilled enough for many other jobs in the private sector. Several GCC governments are not willing or able to provide jobs anymore and many private sector companies prefer foreign workers because they are less complicated to deal with and have a reputation of working harder. (Winckler 2009; Fargues 2013)

All GCC states try to counteract these developments. Companies are demanded to have a minimum employment rate of nationals, but these quotas have not been reached yet, and sometimes they do not get enough applications from locals. (Hertog 2012) There are also reports of formally employed nationals who are complemented by another, often Asian, employee who does the work the national cannot or will not do.[2] On the other hand, all GCC governments show a real interest in improving their educational systems and in making it accessible and affordable for all their nationals. Within a few decades, the GCC states have achieved very high rates of literacy and school enrollment. The quality of education often leaves a lot to be desired and it seems that memorization is still frequently used as a teaching method. Bristol-Rhys sees one of the reasons for this in the provenance of the teachers. (Bristol-Rhys 2008 and interview with author 2009)

A few decades ago gifted students were sent abroad to study but now many Gulf States have brought branches of the world's best universities to the Gulf. As in the economy, the expertise is imported from abroad, creating dependence. One might even argue that the import of foreign expertise and technology becomes a habit that nips the development of local expertise in the bud.

What is more, what students learn at Western influenced business schools may not help them to come to terms with the local business environment. Many young professionals feel that their expertise is not heard when business structures are highly autocratic and marked by top down processes only. (Thompson 2015)

Also on the practical level, the "predicaments" connected with modern education are obvious. Recently, the King Abdul Aziz University of Science and Technology (KAUST) was inaugurated. On its campus, about 80 km in the north of Jiddah, women and men can mingle freely, at least for Saudi standards. In other regions of the country, however, people could still be arrested if they did the same.[3] Islamism, careful to be separated from Islam, is another issue in the GCC states, as in all other Islamic societies. In all Gulf societies, Islamist groupings survey the development process and criticize the government for what they consider to be "unislamic" and corrupt.

Terrorism is also an issue. Gulf nationals have played a sad role in international terror and, given that the GCC states have rather good ties with the USA, the governments are potential targets. Nevertheless, there have been relatively few terrorist attacks in the GCC states. It is probable that GCC governments pay for their security by turning a blind eye to terrorists' transactions in their country or by indirectly supporting their activities.[4] Terrorists could use the mostly liberal economies and financial systems in Dubai or Bahrain and could benefit from this situation, having little interest in changing the *status quo*. However, in the case of Saudi Arabia, not even the large payments the country makes to mosques and Islamic institutions all over the world seem to be sufficient for the bail-out. Saudi Arabia has more than once been the target of terrorist attacks and faces a strong Islamist opposition within the country. These issues are of paramount importance for the perception of country risk, e.g., by potential foreign or local investors. High political risk, which can stem from Islamist or any other kind of opposition, can make low economic or financial risk meaningless, as recent developments during the so-called Arab Spring have shown. (Ramady 2013)

Further important features of the Gulf States are tribal links. All ruling families can trace back their origins to tribes. Usually, there is even a superseding structure like a confederation of tribes, such as the *Banī 'Utbah*. Also clans can arise, for example, as followers of one leading figure within a tribe. Given that polygamy is very widespread in the Gulf region, the sheer amount of children a man can have leads to "substructures" in the family. Sons are sometimes regrouped according to the mother, such as the *Banī Fatima* among Sheikh Zayed's sons in Abu Dhabi. (Davidson 2006)

In addition to sometimes-complicated inner structures, Arab tribes in the Gulf consider themselves as the "original" inhabitants of the Gulf, often looking down on migrants, especially if these are Shi'ites or Indians or Pakistanis. The latter often lack strong kinship ties, which are of vital importance: These ties are reflected in patron and client and family relations within the economy. Private businesses are usually family businesses, positions are given to family members in an apparently vast informal network that exists, especially when it comes to the distribution of tender procedures. The ruling families themselves are the most striking examples of the importance of family ties. Apart from the ruler and the heir apparent, a majority of ministerial positions are usually given to family members, and leading positions in parastatal companies are also appropriate "gifts" for family members. The ruling families themselves have become so vast that their number amounts to a considerable part of the entire population. Furthermore, there are tribes which are deemed sufficiently noble to intermarry with the "royal lineage".

In a study of the "Gulf Family", Ali Al-Tarrah states:

> The main characteristic of Gulf society is that the socioeconomic structure revolves around kinship, which results in a lack of distinction between private interests and public ownership. Overall, the family constitutes the central nerve of economic life, penetrating governmental structures. Society is tightly controlled and dominated by kinship, which includes tribalism, which in turn leads to factionalism. [...] In Arab societies [...] negative aspects of kinship

> have militated against sustainable development. Not encouraged by alternative visions, individual Gulf states have derived their social identity from their tribes, clans or factions. This has made the state a hostage of kinship, thus creating a vicious spiral of misgovernment and a sense of power entitlement in individuals. (Al-Tarrah 2007: 121)

While Al-Tarrah admits that kinship ties could have been used to further development, as in Japan, he regrets that the GCC governments have not taken advantage of kinship in order to promote civic values. Governments have missed the occasion to create new social classes, especially urban middle classes, as drivers of development. Instead of spreading the benefits of an economically active and innovative middle class, the opposite, namely "bedouinizing the city", took place. (Al-Tarrah 2007: 122f) Implicitly, Al-Tarrah argues that the development of a civil society, based on *chosen* ties, worldviews or professions, has been precluded. Instead, sectarianism, mainly the Sunni-Shi'ite cleavage, and working for the best of one's kin, prevail. The Sunni-Shi'ite cleavage, in turn, has become very pronounced since the Iranian revolution in 1979. Today, most Shi'ites in the Gulf States feel that they are discriminated against. In Bahrain, for example, they report that they do not have access to higher posts even if they qualify for them. These situations of course lead to dissatisfaction of important segments of the population with the Sunni ruling families. (Lulu 2011) Eventually, the results have been turmoil and instability, both in Bahrain and in Saudi Arabia.

Today, all GCC governments are faced with a demand for more participation. A century ago, a ruler could personally be approached by his people, although today this is no longer possible to the same extent. The distribution of rents is mostly mentioned as a means to depoliticize the population. The results are neo-patrimonial regimes which enter into a patron and client relationship with their "subjects", rather than a relationship of election and accountability. (Eisenstadt 1973; Erdmann/Engel 2006) There is certainly some truth in this, but it has to be noted that this ruling bargain has not been entirely successful. Even in the wealthiest GCC countries, liberalization and democratization have been demanded by different layers of society. In the interwar period and in times of Arab nationalism, these groups were the merchants, businessmen, and some pan-Arab activists. Today, demand for more participation comes from both conservative Islamic circles, or Islamists, and liberals. The governments, reluctant to install a full-fledged parliament, usually refer to "*shura*" or "advice" as the Islamic form of democracy. Meanwhile, all GCC states have *shura* councils or advisory councils, sometimes with appointed members only, sometimes with a percentage (50 percent usually) of elected members. And despite these councils, all GCC states – with the exceptions of Kuwait and Bahrain, at some point in their history – can still be classified as "traditional monarchies". (Hadenius/Teorell 2007) This is an important commonality which is at the basis of the Gulf Cooperation Council: before its foundation, there were negotiations with the equally oil-rich Arab Gulf State of Iraq, but as a revolutionary system and party to many conflicts in the region, it was not invited to become a founding member. Most recently, the GCC envisioned an enlargement: the two remaining monarchies in the Middle East, Jordan and Morocco, made a bid to join the ranks. (Toumi 2011) An act of monarchical solidarity in times of revolution? Hitherto, no consensus has been reached on the potential new members. (Reuters 2011)

The above mentioned features can be found in all GCC states, but to different degrees. In the following descriptions of the countries in question, I will come back to these issues in order to make clear how a gradual difference came into being and how it affects the economy.

2.2 The Six GCC States

The Economy of Bahrain

Bahrain's modern history was shaped by a comparatively strong British influence and the early discovery (1931) and production of oil. (Birks/Sinclair 1979: 40-43) Having lived mainly from the pearl and trade industries, the ruling family, the al-Khalifa, were encouraged by British advisors to build up a modern administration even before oil income poured forth. Earlier than other Gulf States, Bahrain had a printing press and a cinema, and education reached more than just a small elite. As early as in 1892, Bahrain had a non-*koranic* school and not for boys but for girls.

When oil was discovered and revenues poured in, the foundations for a modern state had already been established. There was already an administration, several hospitals and an educational system, even if quite rudimentary. Most importantly, Bahraini workers found jobs in the oil industry and were prepared to take senior positions. This fact may have contributed to the emergence of something approaching a working class in Bahrain.

However, when the oil-infrastructure had been built and the engineers left, the first depression affected Bahrain. Internationally, this coincided with the decline in the pearl trade and the Great Depression. Already before the war, the skilled Bahrainis especially experienced unemployment, and the channels they had used were no longer possible as the Sheikh could not be approached directly by his population anymore. The various problems led to a movement demanding strikes and more participation. The workers in the oil industry demanded more rights and fewer foreigners. Just before the situation escalated, Sheikh Hamad decided to crush down the movement. Workers of the Bahrain Petroleum Company (BAPCO) who went on strike were laid off.

It was clear from the beginning that Bahrain's oil reserves would not last very long. As Birks and Sinclair put it, Bahrain already prepared for its oil-poor future before oil was discovered. (Birks/Sinclair 1979: 45) A part of this strategy was to add at least a minimum of value before export: Bahrain did not export raw oil, but built a refinery and exported refined oil. BAPCO was founded in 1929 by Standard Oil, and in 1936 its huge refinery opened. Although all GCC states built refineries, mostly this would be decades later. Bahrain's oil production peaked around 1970, and when its own reserves declined, Bahrain started to refine oil from Saudi Arabia. Bahrain also made several wise decisions; for example, exporting its oil to Asia when there was a total lack of refineries. Today, the majority of shares of BAPCO are owned by the state, but there have been mergers with several other companies. British Petroleum (BP) and Caltex are shareholders. To produce natural liquid gas, the Bahrain National Gas Company (BANAGAS) was founded in 1979. It processes natural gas, but only for local use (power, industry, oil production) and not for export. (US Energy Information Administration 2013)

While Bahrain invested a share of its revenues abroad, the government put a clear focus on local development and the creation of job opportunities for its population. As late as in 2006, the Bahrain *Mumtalakat Holding Company* was founded as Bahrain's biggest sovereign wealth fund, and also for this fund's portfolio, the local economy has been weighted heavily. (Mumtalakat 2014) The fund is relatively transparent (8 out of 10 points on the Linaburg-Maduell Transparency Index). (Sovereign Wealth Fund Institute 2013)

As energy intensive branches, heavy investments were made in the aluminium and steel industries. Founded in 1968, Aluminium Bahrain (ALBA) is still one of the biggest aluminium smelters worldwide, with a yearly output of 870,000 metric tons. The amount of output allowed the opportune creation of an aluminium downstream industry: Bahrain Atomisers International produces aluminium powder, the Aluminium Extrusion Company (BALEXCO) processes aluminium further, Middle East Aluminium Cable (MIDAL) produces cables, and there is also an aluminium rolling mill (Gulf Aluminium Rolling Mill Company, GARMCO). However, since all these industries are energy intense, Bahrain depletes its gas reserves increasingly faster and may become a net importer of NLG soon. Already, Bahrain depends upon the good will of Saudi Arabia, who shares revenues from the Abu Safa oilfield, upon which Bahrain put a claim on until 1958, and exports raw oil to Bahrain for refinement. (Zahlan 1998: 76; Kubursi 1984; Kvande 2012: 218-220; Mills 2012)

An early diversification project has been the shipbuilding and amending company ASRY. While considerably enlarged in 1977, it has to worry about its competitiveness today. Furthermore, small and medium companies have been built up and encouraged by the Bahraini state. Metallurgy and textiles are especially important.

Because of its relatively liberal society and, for example, the sale of alcohol, Bahrain has become a center for tourists from other Gulf States. Saudis often escape the narrow confines of their country to spend a few days in Bahrain. Expats sometimes prefer to live in Manama, even if their jobs are actually in Saudi Arabia. (Kubursi 1984)

When in 1973 the oil price rocketed, Bahrain's reserves had already declined. However, since Bahrainis had been aware of their limited resources, they felt the need to diversify very early. Her own resources declining, Bahrain grasped the occasion to make use of the wealth that was being accumulated elsewhere: banking in Bahrain started early in the 20th century. (Gerakis/Roncesvalles 1983) By 1960, four commercial banks operated in Bahrain, among them the Bank of Bahrain (1957) which was to become the National Bank of Bahrain.

Hence, there was already a cadre of expertise and fifteen more commercial banks were founded between 1969 and 1977. In 1973, the new Bahrain Monetary Agency took the decision to host offshore banking units: foreign banks now had the opportunity to benefit from the accumulated surpluses in the region, and Bahrain became the financial center of the Arab Gulf. Bahrain also benefitted from the civil war in Lebanon where what had been hitherto the financial hub of the Middle East crashed. Decision makers in Bahrain were again lucky and clever enough to grasp the opportunity. Bahrain neither missed the train of Islamic Banking: In 1979, the Bahrain Islamic Bank was founded and 27 other Islamic financial institutions followed soon thereafter. Today, around 400 banks, both conventional and Islamic banks, operate in Bahrain, and the Central Bank of Bahrain is usually attested to as a sound supervisor of financial activities. There is also an Accounting and Auditing Organization for Islamic Financial Institutions, providing guidelines for Islamic Banking – so far a unique institution in the Persian Gulf States. Today, the financial sector contributes around 27 percent of the GDP. This development was also possible because of the relatively high share of skilled workers. In 1981, the Bahrain Institute of Banking and Finance was founded to train Bahrainis and also many foreigners for jobs in banking and finance. (Alkhozai 2008: 145; Grigorian/Manole 2005:10) Many Bahrainis took jobs in the new banking sector, and this gave rise to a middle class that cannot be found in this dimension elsewhere in the Gulf. More than twice as many Bahrainis work in the private than in the public sectors. (Hertog 2013: 49) Even in the banking sector around 70 percent of the employees are Bahrainis. On the other hand, unemployment (around 15 percent) is a problem. Given that Bahrain never had the option of endless oil reserves, rent distribution is less important than in other Gulf States. In 2007, the Bahraini government even introduced a very low income tax in order to fund a new unemployment scheme, despite massive opposition from the society. (Khaleej Times/DPA 2007)

Nevertheless, the boom of the 1970s attracted many foreign workers although Bahrainis remained the majority. This might have changed due to the recent oil price peak. The 2010 census estimated that around 54 percent of the population were foreigners. (Kingdom of Bahrain/Central Informatics Organisation 2010) Recently, a McKinsey (Zogby 2007: 32) survey looked into the "middle classes" in the UAE, Saudi Arabia, and Bahrain. Compared to Saudi and UAE citizens, Bahrainis were less confident, and complained about less leisure time, little recognition for their job, and low salaries. More respondents from Bahrain were afraid to lose their job or had lost it recently. While at first sight this seems to be a rather bleak view on the Bahraini middle class, it could actually be a sign of a more meritocratic economy: apparently, Bahrainis share the fears of middle classes all over the world. They do not have the promise by the government that their job is guaranteed, they may sometimes need to work extra time or take jobs other Gulf citizens would not take. As such, the Bahraini responses may reflect the fact that the Bahraini government cannot afford rentier mentality among its citizens. This is confirmed by another McKinsey study on education in the GCC: In Bahrain, Bahrainis outnumber expatriates in medium and low skilled jobs. (McKinsey 2007)

Bahrain is known for its liberal business climate. The Heritage Foundation rated Bahrain as the freest economy in the Middle East and North Africa. It ranks just after the US and Ireland on rank 12, while the next Middle Eastern country, Qatar, only achieved rank 27. (The Heritage Foundation 2013) In 2001, the new Commercial Companies Law was promulgated. This law provides for several business structures, and some can be entirely owned by foreigners. Nevertheless, several business activities remained limited to Bahrainis, such as publishing or oil product supply. The liberal business climate has attracted foreign direct investment, rising from 0.2 billion in 2003 to 1.1 billion in 2006, and the World Bank's "Doing Business" index asserts Bahrain a business friendly environment: Bahrain was at time on rank 20 but the rankings change considerably from year to year. In 2014, it only reached rank 53. (Khan 2006) Among the single rating categories, enforcing contracts, and interestingly getting credit, are the most problematic aspects of "Doing Business". (WB Doing Business 2009; 2010) In 2004, Bahrain signed a Free Trade Agreement with the USA which came into effect in 2006. Among the GCC states, Bahrain was the first country to sign such an agreement.

Bahrain's governance ratings also stand out (see table 2). Voice & Accountability are rated low for all GCC states, while the political instability, probably due to the Shi'ite-Sunni divide and the strong opposition activities for more participation, has not deterred investors. Bahrain stands out for its high regulatory quality.

***Table 2* Bahrain's Governance Ratings**

Year	Voice & Accountability	Political Stability	Government Effectiveness	Regulatory Quality	Rule of Law	Control of Corruption
1996	-1.19	-0.82	0.387	0.522	-0.1	-0.25
1998	-1.04	-0.21	0.781	0.803	0.497	0.299
2000	-0.92	0.071	0.681	0.912	0.463	0.442
2002	-0.69	0.11	0.63	0.975	0.663	0.821
2003	-0.69	0.208	0.508	0.735	0.623	0.651
2004	-0.58	-0.01	0.589	0.818	0.773	0.562
2005	-0.73	-0.27	0.416	0.729	0.638	0.493
2006	-0.83	-0.38	0.374	0.718	0.508	0.33
2007	-0.84	-0.26	0.449	0.857	0.621	0.371
2008	-0.85	-0.23	0.479	0.797	0.607	0.351
2009	-0.83	-0.09	0.624	0.778	0.508	0.297

Source: World Bank Doing Business (2010).

Today, Bahrain's financial sector, of course, suffers from the financial crisis, but Bahrain could have been hit worse. The Oxford Business Group attests the Central Bank of Bahrain employed a sound and swift counter strategy. (Oxford Business Group 2012) Some analysts even forecast a boost in Islamic finance because of a general return to simple, transparent and "trustworthy" financial products.

In 2007, the oil and gas sector made up only 24.6 percent of GDP and the diversification can be regarded as quite successful. The Herfindahl-Hirschman index for Bahrain, a measure of market concentration, attests the country underwent an improvement from a comparatively low 0.400 in 1995 to a 0.356 in 2013, although it had received higher values in the meantime. (UNCTAD 2014) This measure is not corrected for oil prices fluctuations.

My own calculations assert Bahrain a sound growth performance of its non-oil GDP: on average, it grew 8 percent *per annum*. While Bahrain has not practiced economic planning using 5-year plans as in Saudi Arabia or Oman, the government presented its "Economic Vision 2030" in 2008, addressing several key issues – such as an oversized public sector, unemployment, and a lack of innovation – very bluntly. The paper puts a clear focus on better education, a better business climate, a competitive private sector, and on incentives for innovation in the private sector. A high standard of social security would be maintained, but "Bahraini society in 2030 will be a meritocracy where hard work and talent are rewarded with success". (Bahrain EDB 2008:16) While the "economic long-term visions" have become very popular among GCC governments, Bahrain's plan is the most daring attack on the allocation state. It aims at reducing the role of the state, while the future motor of growth should be a competitive private sector. (Hvidt 2013: 219-221)

Bahrain is the only GCC country where massive protests took place during what was widely called the Arab Spring in 2011. The country has a long history of demands for more political participation and more rights for the Shi'ite majority. Equally long is the history of violent suppressions of these protests and the dissolutions of parliament. In 2011, the government responded to the protests by releasing political prisoners but the main reaction was violent repression: in March 2011, Saudi Arabia sent military forces into Bahrain to restore peace and order. For Saudi Arabia this was probably a pre-emptive step to preclude any spillover effect on its own Shi'ite minority. (Noueihed/Richter 2011) It remains to be seen whether these events will harm the reputation of Bahrain as a rather stable and liberal country.

The Economy of Kuwait

Kuwait's economy before the discovery of oil was marked by pearls, trade, and a viable merchant community that has since become a very powerful political force in the country. (Al-Hijji 2010) The British protectorate in Kuwait ended in 1961, and earlier than its neighbors, the Kuwaiti government was faced with the task of codifying its commercial law. It decided to use the Egyptian (French inspired) law code as a role model. (Amin 1991) As far as personal status law was concerned, it was convenient to keep the Ottoman *Majallah* as the Civil Code. In 1980, the Ottoman *Majallah* was superseded by Kuwait's own Civil Code, while in 1984, a code of Personal Status was promulgated – an exception in the GCC countries. From 1959 to 1964, business related laws were also drafted and promulgated, among them the Kuwaiti Commercial Code in 1961. All these measures reflect the clear determination of the Kuwaiti government to establish a modern and secular legal system. However, around 1980, several laws were amended in order to bring them in accordance with the *sharia*. Even if the provisions of the Commercial Law were not amended, judges increasingly resorted to *sharia* law. (Coulson 1984: 5f)

This early codification of law allowed for other developments: the banking system operated at a high level as early as in the late 1950s. In 1952, the National Bank of Kuwait, up to today an important, entirely Kuwaiti-owned commercial bank, was founded. In 1969, a general Banking Law was promulgated and in the aftermath, the Currency Board, having replaced the Indian *rupee* with the Kuwaiti *dinar*, was turned into the Central Bank of Kuwait. Also, the history of the Kuwait Stock market goes back to 1962.

Since the discovery of oil in 1938, the oil sector has clearly been the most important sector and Kuwait's economy has so far hardly diversified away from oil. Depending on the source, Kuwait's oil reserves are estimated at around 100 billion barrels or approximately 8 percent of the world's reserves. In addition to her own oil fields, Kuwait produces oil from the Divided Zone between Kuwait and Saudi Arabia. In 1934, the concession for Kuwait's oil was given to the Kuwait Oil Company (KOC), which was then jointly owned by the Anglo-Persian Oil Company and the Gulf Oil Corporation. As in other Gulf States, the 1970s saw the nationalization of KOC in 1976. In 1980, the Kuwait National Petroleum Corporation (KNPC) was founded as a state-owned umbrella company for KOC and several other downstream companies in the oil sector. Meanwhile, several downstream projects have been realized and Kuwait refines oil, produces aluminium, petrochemicals, fertilizers and building materials, but the gross of its exports (around 95 percent) still consist of oil. Its oil wealth, however, contrasts with a relative shortage of non-associated gas which is an important feedstock of the energy-intense petrochemical sector and has to be bought at a relatively high price. (Henni 2012)

Apart from the oil sector and related fields, the welfare system and infrastructure were developed, causing a relative weighting of the construction sector and at a very early distributive state: every Kuwaiti, even if he was simultaneously a shop owner or trader, had to have a job in the public sector, often without having any qualifications. In many cases, Kuwaitis were given "guard" jobs and soon, every institution or hospital had more guards pacing the entry halls than it needed. In qualified jobs, Kuwaitis were practically absent. Free health care, job security, education, and infrastructure were ways to distribute income from oil and to distract Kuwaitis from pan-Arab agitation by putting a rift between Kuwaitis and Arab expatriates. Buying off the Kuwaiti population also meant giving them privileges which expatriates, who already accounted for 45 percent of the population in 1957, did not have. As early as in 1955, the British political agent wrote that "amongst all classes of Kuwait the idea is growing that while paying no taxes they are entitled to free or State-subsidized education, health, water, electricity, telephones and other services." (Crystal 1995: 79)

Furthermore, the strong merchant community reached a tacit agreement that they would be granted monopolies in trade, and that the royal family would not compete with the merchants in this field. It also seems that the local merchants did not appreciate the local *production* of consumer goods and were not in favor of import substitution because they make profit by importing goods that are produced abroad. Local production would be competition to their trading activities. These special privileges for the merchants are quite unique in the Gulf. They have meanwhile resulted in a comparatively strong parliament and the inactivity of the government in several realms.

When the Commercial Law was enacted it contained a whole array of privileges for Kuwaitis, particularly provisions limiting the rights of foreign investors and strengthening the role of local sponsors. A few local firms owned by rich merchant families were given government contracts to enrich themselves even further without really contributing anything:

"Government contracts went to a few old, established domestic firms. In practice, the nationals in these firms often contributed neither capital nor management. Purely paper participation proliferated." (Crystal 1995: 90)

The government may thus have deliberately chosen to rely on the oil industry and on overseas investments for income: Kuwait's first sovereign wealth fund had already been founded in 1953. Meanwhile, there are several funds (Reserve for Future Generations, General Reserve Fund) operating under the Kuwait Investment Authority (KIA).[5] Given that Kuwait's oil reserves are immense, there has been no urgency to diversify aggressively and to risk more opposition from the population and the merchants. Moreover, the first attempts to diversify were not very successful and even regarded as a "costly failure" in public opinion. (Owen/Pamuk 1998: 216)

This approach continued when oil prices plummeted in 1986: Kuwait was affected, but since its revenue from foreign investments already outweighed oil revenue, it could have been hit harder.

> By choosing to invest in oil diversification and foreign markets rather than in local development, the rulers had decided to reproduce the rentier nature of the oil economy through investment abroad rather than face the risks, political as well as economic, of local development. This choice has generated a substantial non-oil income, but one which, like oil, continues to flow directly to the state, reinforcing the ruler's oil-generated independence from other elites, but at the cost of dependence on distant financial markets. (Crystal 1995: 96-97)

Thus, the Kuwaiti government's fear of opposition has given rise to an extremely inefficient economic situation. This also became clear during the Suq al-Manakh (Sūq al-Manākh) crisis in 1982. The Suq al-Manakh was an illegal but very popular stock market mostly financed by post-dated exchequers. Its market volume was estimated at around US $93 billion – this was around four times the annual GDP of Kuwait at that time. While there were a few big investors, it seems that speculation had become a popular sport of the average Kuwaiti. In 1982, the marked crashed, triggered by someone who actually tried to cash one of the post-dated exchequers. This crash resulted in the bankruptcy of six out of seven operating banks in Kuwait. The government, however, did not only turn a blind eye to this development, it reacted by buying shares on the market to maintain the price level. Nevertheless, many Kuwaitis had accumulated debts. These nationals were supported by a "Difficult Debts Program" and eventually all Kuwaitis had their debts written off after the Iraqi invasion. The Kuwait government spared no effort to avoid conflict, and bailed out its citizens. (Fitch 2010; Azzam 1983)

When Iraq invaded Kuwait, Kuwait's foreign assets had to be partly liquidated and Kuwait even had to borrow money. (Sadowski 1997) Already before the war, low oil prices had caused budget deficits. Now, it even came worse: during the months of the Iraqi occupation until Desert Storm, much of Kuwait's infrastructure was destroyed, especially in the oil sector. When the Iraqi army withdrew in January 1991, it set oil wells on fire, burning around 2 percent of Kuwait's estimated reserves. In short, the costs of the war were disastrous and the reconstruction process was severely hampered by corruption.[6]

Instead of cutting back subsidies for the population on this occasion, the Kuwaiti government even paid extra: families who had stayed in Kuwait during the war had their electricity or water bills paid and received a lump sum of US $1750. Apparently, the priority was to restore confidence in the Kuwaiti government.

It is thus not astonishing that the Kuwaiti government was eager to raise its revenues, mainly by a rapid reconstruction of the oil sector and by expanding production. Karen Pfeifer sums up the economic activities in Kuwait after the war as follows: "For the decade 1991-2001, Kuwait's main economic growth strategy was simply to pump and sell as much oil as possible." (Pfeifer 2002: 12) Kuwait even increased its reserves. The Kuwaiti-Iraqi borders were finally drawn without Iraqi consent. (Pfeifer 2002: 10-13) The Rumaila oil-field, which had only *bordered* on Kuwait before the war, was now partly *in* Kuwait – even if only for the very southern peak. Since then, Kuwait has tried to increase its production capacity and considered economic reforms. To raise revenues, the Kuwaiti government even thought of privatizing part of the oil industry. In an attempt to open up the Kuwaiti oil industry for international investors in 1997, foreign oil companies were considered potential partners for the construction of the new infrastructure for oil production. This was introduced as "Project Kuwait". There was opposition to this project in parliament however.[7] There are also projects to modernize the three Kuwaiti refineries, as well as projects to build new power stations, although most projects have been impeded or delayed by the National Assembly, which had become stronger due to the concessions the Sheikh had to make after the Iraqi invasion.

The same holds true for infrastructure projects. The main goal of the National Assembly, it seems, is to defend the privileges various groups have obtained over the last fifty years.

There is still an absence of Kuwaitis in private sector activities. Ninety percent of the Kuwaiti labor force work in the public sector and are better paid than foreign workers. It is a declared aim of the government to raise the share of Kuwaitis in other sectors, if necessary by quotas. It remains a problem that Kuwaitis still lack the skills or the will to work in certain jobs as long as they can have an easier, more lucrative job in the public administration. While the official unemployment rate stood at an estimated 2.2 percent in 2004, this rate would be much higher if unnecessary jobs in the public sector were eliminated. Furthermore, subsidies for water and electricity lead to over-consumption by the inhabitants. This practice also impedes projects in renewable energies, which would hardly produce energy at such a low price level. Obviously, the government still shies away from changing what Kuwaitis consider their right. Nonetheless, bad working conditions of foreign workers and their lower pay have meanwhile become a frequent subject of Kuwaiti media attention. And very few reforms target the typical sectors for economic diversification: there is, for example, a touristic project on the small island of Falaika.

Foreign direct investment has not become an important issue yet.[8] Despite a quite modern business law, several restrictions were problematic: firms had to be 51 percent owned by a Kuwaiti, and local sponsors were necessary for most business activities. There are laws obliging the employer to preferance Kuwaiti employees, and taxes imposed on foreign companies were at 55 percent until recently (now reduced to 25 percent). In addition, Kuwaiti firms were preferred for tenders in public projects. This is why, when the laws on foreign companies were changed in 2003, most companies preferred to stay "Kuwaiti". Usually, foreign companies have only commercial settlements in Kuwait. Furthermore, Kuwait has done nothing to reduce corruption and administrative barriers for foreigners. These problems are also reflected in the country's "Doing Business" ratings which have dropped to rank 61 in 2010 and even to rank 86 in 2014. Starting a business is obviously the most problematic issue (only rank 137 in 2010), but also contract enforcement and across border trading are seen as problematic. The only bright spot in Kuwait's ranking seem to be the low taxes and the relatively easy employment of workers, which, to be sure, is mostly to the detriment of labor migrants. (World Bank 2010) In 2013, the foreign investment law of 2001 – which had contained a huge negative list – was revised, opening up more sectors to foreigners. Additonally, the Kuwait Direct Investment Promotion Authority has been created in order to promote the effects of the 2001 law. These effects remain to be seen. (Shabana/Salah 2013)

In general, Kuwait has been declared as a high potential for development but observers agree that this potential cannot be implemented under the current political conditions. (GTAI 2013) This is also reflected in the country's low ratings for regulatory quality and government effectiveness (see table 3). It seems that Kuwait, without being a truly democratic country, also lacks the "embedded autonomy" required for economically effective measures.

Table 3 **Kuwait's Governance Ratings**

Year	Voice & Accountability	Political Stability	Government Effectiveness	Regulatory Quality	Rule of Law	Control of Corruption
1996	-0.47	0.006	0.311	-0.02	0.627	0.539
1998	-0.27	0.353	0.138	-0.02	0.738	0.989
2000	-0.28	0.606	0.089	-0.05	0.708	0.985
2002	-0.38	-0.14	0.145	0.326	0.677	1.179
2003	-0.44	0.083	0.265	0.291	0.615	1.002
2004	-0.36	0.158	0.319	0.514	0.65	0.975
2005	-0.31	0.217	0.423	0.547	0.706	0.863
2006	-0.39	0.374	0.333	0.455	0.658	0.797
2007	-0.52	0.556	0.167	0.309	0.673	0.508
2008	-0.56	0.505	0.06	0.219	0.644	0.552
2009	-0.54	0.418	0.206	0.197	0.591	0.424

Source: World Bank 2010.

In 2007, the Central Bank of Kuwait published a Kuwait Vision 2020 (Central Bank of Kuwait 2007) and in 2010, the Kuwaiti parliament approved a four-year development plan, the first since 1986. This plan is supposed to be the first of six plans leading up to 2035. Furthermore, the negative repercussions of the financial crisis 2008 have caused the government to commission the consultancy firm of Tony Blair Associates for a development plan for the following 20 years. This "Blair Report", otherwise named "Kuwait Vision 2035" (Kuwait News 2010), can be considered as the background of the shorter development plans. This report points to the usual weaknesses of rentier societies, but also to the unproductive relationship of parliament and government. Apparently, this lack of consensus has also precluded earlier development planning. (Crystal 1995: 95ff.)

To sum up, the case of Kuwait is one of development being extremely hampered by political considerations and the effects of the rentier state. According to my calculations, Kuwait's non-oil sector "grew" on average -2.29 percent. In 1995, UCTAD's concentration index assigned Kuwait a 0.610 which points to a comparatively highly concentrated export portfolio. Apart from a few low oil-price years when Kuwait ranked slightly below 0.6, the concentration only grew further and stood at 0.677 in 2008 and even 0.719 in 2013. (UNCTAD 2014)

The Economy of Oman

Like the country itself, Oman's economy was divided until 1955. While the coastal region was an important trading place where several foreign minorities lived, the interior (the Ibadi

Imamat) was a subsistence economy. This was true even after the British had forbidden slavery and after competition from India had brought down Omani textile production. What remained were the pearl trade, agriculture, fishery, date cultivation, and handicrafts. There was something like a banking system, and small shop-keepers and bazaars. Import and export continued and the most important trading partner was India. If interior Oman traded, it usually avoided the towns of coastal Oman (Muscat and Maṭraḥ) and traded via Sohar (Ṣoḥār). (Speece 1989)

Inner Oman is famous for its traditional irrigation systems called *'aflāj (*from singular *falaj).* These highly sophisticated systems that tapped ground water allowed the population to pursue successful agricultural projects despite difficult circumstances. Thus agricultural production is deeply rooted in Omani society, even if only a very small part of Oman's land has been used for agriculture. One could even argue that the elaborate distribution and sale of shares of irrigation water among the population caused a special social fabric otherwise different from nomadic societies. Not much economic development took place before Sultan Qaboos came to power. Oil was only discovered in 1964 and has been produced since 1967 by Petroleum Development Oman (PDO). The Omani state owns 60 percent of PDO, and 34 percent are held by Dutch Shell.

Oman's oil reserves are a subject for speculation. They have always been estimated as rather modest, but the geological situation of Oman makes these estimates quite difficult. Oil production is more complicated and expensive than, for example, in Kuwait. According to older estimates by Dutch Shell, oil reserves were around five billion barrels, but in 2004, Shell revealed that it may have overstated the amount of oil in Oman. However, Oman keeps discovering new reserves. Several times, Oman's oil production declined, probably because the current location was being depleted. As for the current outlook, Omani oil could be depleted in about 20 years. Oman also has some natural gas, which has been produced since 2000. It is mainly used for electricity and the desalination of water. (Katz 2004: 1f) Like Bahrain, Oman has tried to at least partly add some value to its raw oil by refining it before export, and a refinery was established in 1982. Today, the Oman Refinery Company is part of Oman Refineries and Petrochemicals Company LLC. (ORPIC 2013) Also some industrialization has taken place. There are copper, cement, marble, gypsum, and limestone industries, but Oman has avoided the huge and prestigious plants of petrochemicals production, which would have suffered from a strong regional competition. (Whelan 1987: 141f)

Oman also utilizes its "historical" advantage in agriculture and encourages private investment into agriculture by offering soft loans. The main target is the local population. To this end, the *Oman Bank for Agriculture and Fisheries* was founded in 1981. Also the fishing sector is promising and the Omani government supports fishermen by facilitating their purchases of modern equipment. This traditional sector continues to attract local Omanis.

Despite its importance for the employment of nationals, agriculture is not without flaws. The fact that the locals do not depend more on agriculture has caused an irresponsible use of sources. New technologies have led to sinking ground water tables and a higher salinity of wells. (Grandmaison 1984; Gabriel 1990) Furthermore, farmers often lack the expertise to grow the products demanded by a modernized and more demanding society, and growing them would disturb the equilibrium of irrigation. Hence, Oman experienced a flight of the rural population to the cities, and oftentimes small agricultural enterprises have only survived due to labor remittances gained by male household members abroad. In 1990, the Ministry of Water Resources was founded to confront these problems and to establish water saving methods. (Dutton 1987)

Before the coup of 1970, little development took place under Sultan Said, widely believed to have been reluctant to develop his country. However, this picture of a despotic ruler averse to any kind of modernization has been increasingly eradicated (Rabi 2011). Nonetheless, it is safe to say that systematic economic development only started after 1970 following five-year development plans. The first one was launched in 1976 and was embedded into a larger vision reaching from 1970 to 1995. From 1995 onwards, the framework for planning is the "Vision 2020". These plans are prepared by the Ministry of National Economy. Their main goal is economic diversification and a reduction of the oil sector's contribution to GDP. Even if Oman also develops its gas sector, the main focus is on tourism, local human capital, and a more diversified production base.

Oman also took several steps to open its markets for private investors, including foreigners. While at least 51 percent of a company must still be owned by an Omani, there have nevertheless been incentives such as large tax holidays. Additionally, Oman is about to create a free zone, operated by the Jebel Ali Free Zone Authority (Dubai), where companies may be 100 percent in foreign hands. Apparently, this strategy is successful. In recent years, Oman has experienced rising foreign direct investments, both in absolute terms and as a percentage of GDP. (ESCWA 2008: 6; ESCWA 2009: 3) Oman also successfully attracts tourism and can benefit from a rich cultural heritage. Nevertheless, in 2010 Oman was only on rank 65 of the "Doing Business" index. In addition to the notorious problems of contract enforcement, there seem to be bottlenecks regarding credit and construction permits, and trade across borders seem to be problematic. (WB Doing Business 2010)

According to the World Bank, Oman has a sound institutional setting, as table 4 shows. It combines high political stability and rule of law with a high regulatory quality and government effectiveness, which is probably based on the quasi absolute power of Sultan Qaboos and his clear determination to develop his country.

It is also worth looking at the Omani Banking sector. In 1974, the Central Bank of Oman, which had been the Oman Currency Board before, started operations. It fulfills all the task of a normal central bank. There are both local and foreign banks operating in Oman. Among the local banks, three are specialized, financing local development and helping nationals (and GCC citizens) to purchase or build houses. (The Oman Housing Bank, OHB) However, during the entire period of this study, there was no Islamic

Table 4 Oman's Governance Ratings

Year	Voice & Accountability	Political Stability	Government Effectiveness	Regulatory Quality	Rule of Law	Control of Corruption
1996	-1.02	0.47	0.797	0.119	0.727	-0.25
1998	-0.65	0.693	0.519	0.136	0.756	0.674
2000	-0.66	0.862	0.514	0.2	0.7	0.646
2002	-0.74	0.781	0.548	0.762	0.624	0.874
2003	-0.92	0.805	0.64	0.708	0.606	0.444
2004	-0.63	0.829	0.596	0.712	0.674	0.644
2005	-0.73	0.783	0.504	0.705	0.527	0.464
2006	-0.95	0.778	0.496	0.817	0.524	0.458
2007	-1.06	0.851	0.43	0.696	0.576	0.4
2008	-1.06	0.886	0.537	0.796	0.77	0.568
2009	-1.08	0.812	0.645	0.661	0.683	0.479

Source: World Bank 2010.

banking in Oman despite pressure from banks and the public to open the market to *shari'ah* compliant products. In 2007, Hammood Sangour al-Zadjali, executive of the Central Bank of Oman, simply stated "We shall not allow it. We believe that banks should be universal. We won't allow specific banks." (MEED 2007) While it seemed that Qaboos wanted to leave this growing market to his neighbors, protests in 2011 finally led to a legalization of Islamic banking in Oman.

Since the 1970s, Qaboos put a special focus on providing the population with education and health care and other social services. These are major budget items and they have, as elsewhere, led to a growing administration. Most Omanis used to work in the rising public sector or the administration, while comparatively few worked in the private sector. In many cases, however, Omanis lacked the necessary skills for both the private and the public sector, which is not astonishing because educational efforts started very late. However, Oman's human resource development strategy was commended by analysts. In 1987, Birks and Sinclair assert Oman's education strategy "a strong and imaginative start". (Birks/Sinclair 1987: 167)

Moreover, despite the late development, Oman was the first[9] to launch projects to "nationalize" the labor force and to raise the share of Omanis in the private sector, mostly by educational efforts but also by quotas and preferential treatment of companies reaching these quotas. When judging the result of these efforts, one has to keep in mind that Oman started from almost zero in 1970 and compare with other Gulf countries, where these efforts had little effect. Katz deplored that, in 2004, only 10 percent of private sector employees were Omanis.

(Katz 2004: 2) However, in other Gulf States, nationals only make up for one percent, despite educational efforts having started earlier. Omanis today work in jobs other Gulf citizens would rather avoid. There are Omani nurses, teachers and drivers, and some work in hotels and industry. The focus on vocational training has had a modest but real Omanization effect. According to the 2010 Census, 816,143 expatriates live in Oman, around 270,000 more than in 2003. (Oman Census 2010: 5) With a total population of approximately 2.8 million, this compares well to other GCC states. Obviously, not every Omani has the idea that the government needs to provide him or her with a job, even if this may be true for the better-skilled Omanis. The less-skilled even go to other Gulf countries to work there in hotels or as drivers. For Gulf standards, they have the reputation of being hard workers.

Oman's diversification efforts have been comparatively successful so far: after Bahrain and Dubai, it is considered the third most diversified GCC state. From a very high export concentration in 1995 (0.748), it reached 0.621 in 2008 and 0.591 in 2013 despite higher oil prices. I calculated an average non-oil growth rate of 2.5 percent. Oil is still the major source of revenue although it is uncertain if non-oil projects will be profitable once no longer subsidized by oil revenues.

Despite the recent establishment of councils with appointed members, the rule of Sultan Qaboos is an absolute monarchy. Contrary to other Arab Gulf states, all important positions are concentrated in his person. (Peterson 2013: 319f) Michael Herb argues in his book *All in the Family* that Middle Eastern rulers who do not include their family in their decisions are more likely to be overthrown. Qaboos has no children. (Herb 1999: 180) It appears that he wants his family to decide who will succeed him after his death. In case his family does not agree on a candidate, he is said to have given a sealed letter with two names (ranked) to the Minister of Defense. (Peterson 2012) The stability of Oman seems to depend on the very person of its Sultan, and prospects of leadership after him remain vague.

The Economy of Qatar

Due to its immense natural gas reserves, Qatar is today the richest of all the six GCC states. With a *per capita* GDP of around US $76,000, Qatar is among the wealthiest countries if not *the* wealthiest country in the world. In the 1970s, Qatar saw a tremendous rise of GDP, but today's wealth is mainly a result of Qatar's natural gas reserves. While Qatar's oil wells are expected to be dry in a few decades, its natural gas reserves are the third largest in the world, probably amounting to around 900 trillion cubic feet.

Qatar has a long history of disputes and coups in its divided ruling family, the Al-Thani, but contrary to Kuwait there is little opposition from any traditional merchant elite. (Crystal 1995: 118) Before the discovery of oil, the Qataris lived from fishing, pearl-diving and some agriculture, although like other Gulf States, Qatar was hit hard by the depression and subsequent decline of demand for natural pearls in the interwar period. The concession to explore oil was granted in 1935 to the Anglo-Persian Oil Company. In 1940, oil was found in commercial quantities and the oil sector soon became the most important economic sector although the oil reserves were not deemed really important. The PDQL did not even have a representative in Qatar. This was probably one of the reasons why, in 1952, a rather unproductive pattern of negotiations between the PDQL and the Sheikh for the shares emerged. Objects of these negotiations included a 50/50 share of revenue, preferential employment for Qataris, meals for workers, and the share of the pay servants had a right to keep for themselves. There were frequent strikes by the oil workers, which the Sheikh (although Emir is more correct, for Qatar) used to boost his position. The representatives of the PDQL usually were arrogant and adamant, whereas only a well-meant gesture would have been required to improve relations with the Sheikh and the workers. In 1976, the PDQL was nationalized.

Oil workers included British seniors, foreigners (Indians), and also Qataris – both servants and free men. The difference with other Gulf countries, Kuwait particularly, was that Qataris took these jobs themselves, not via any government effort to create jobs for the people. The "rentier system" only started later. This may be due to the fact that Qatar was one of the smallest oil exporters. In OPEC, it was on a level with Ecuador and Gabon. Furthermore, given that Qatar is such a young entity, it is unclear whether such a state could survive independently.

For a long time, the rulers of Qatar seemed to be mainly interested in securing their private income from oil while at the same time containing the demands for allowances to family members. Public and private finances were hardly separated. (Rathmell/Schulze 2000: 59) Nevertheless, the first revenues from oil allowed for modest development. After the first shipment of oil had left the country in 1949, the government founded the first schools and expanded the health care system. It is noteworthy that the population of Qatar mostly adheres to the Hanbali (ḥanbalīy) school of Islamic law, considered the strictest interpretation of Islam. Historically, there was also a strong Wahhabi influence. More than in the other small Gulf States, the educational focus was on religion and the Arabic language. (Sadik 1972) In 1973, the public Qatar University was founded, while today, several Western universities have established Qatari branches.[10] There has also been a rise in English-language international schools, where teaching is in English only.

The real efforts to diversify the economy started in the 1970s, still ignorant of the tremendous wealth in natural gas resources. In 1965, the Qatar National Cement Company was created, and later on, ventures in fertilizer production, steel and petrochemicals met with mixed success. Qatar Steel, formed in 1974, had massive problems at the beginning and made its first profit only in 1988. Other public companies show only low profitability despite the comparative advantage of cheap energy. The Qatari government also invested in agriculture and livestock, but these sectors are only minor contributors to the GDP due to the country's climate and lack of arable land. Despite massive investment by the government there has always been a focus on private enterprise. Early on, Qatar tried to attract foreign investors and technology and facilitate private business. Qatar also offered an array of subsidies, tax exemptions, and protective tariffs. From 1989 on, the industrialization process was monitored by the new Ministry of Industry, while in 1993, a specialized bank was establish to improve credit possibilities for industrial companies.

Given its limited oil reserves, the Qatari government even implemented production ceilings in the 1970s and 1980s. It also contracted foreign companies to make exploration more efficient and to find new reserves. And indeed, between 1995 and 2000, Qatar's oil reserves tripled. Nevertheless, it was clear that Qatar had to develop her immense natural gas reserves of the North Field in the long run. Already in the 1980s, local energy consumption had been increasingly met by natural gas. Gas exports, however, required large investments into infrastructure and new plants. This was increasingly difficult in the 1990s when revenue from oil had become very low. After the budget surpluses in the early 1980s, Qatar now had to return to domestic and foreign borrowing as well as to the liquidation of foreign assets. GDP shrank due to the heavy reliance on oil and gas. Under Emir Hamad, there were even some cutbacks on welfare: subsidies for food and daily utilities were lowered and nominal fees for education and health care were introduced. Hamad also took steps towards privatization; for example, by establishing the Doha Stock Market in 1995, by strengthening the role of the Chamber of Commerce, and replacing the Ministry of Water and Electricity with a corporation. By cutting back subsidies and budget austerity, Qatar was able to reduce debt significantly.

Meanwhile, the harvest from expensive investments into natural liquid gas plants and infrastructure can be collected: while Qatar only produced around 29.5 billion cubic meters of natural gas in 2002 (and much less in the 1990s), the production rose to 145.2 billion cubic meters in 2011. (BP 2013: 22) Qatar exports to Europe and the USA, but especially to China and other Asian countries. Nevertheless, Qatar does not want to solely rely upon gas, and from 2004 to 2008 Qatar's non-oil growth performance improved. There were plans to invest in tourism, specifically business travel and international conferences. Qatar aims at attracting the *crème de la crème* of tourism by hosting cultural events, museums, and sport events. To this end, the Qatar Tourism Authority was formed in 2000. In 2022, the FIFA World Cup is supposed to take place but the climate in Qatar, rumours of corruption during the bidding process, and reports of the bad working conditions of foreigners have given rise to demands to go back on this decision.

For expatriate workers, Qatar really seems to be a rather unloved destination. Blogs of expatriates in Qatar abound in reports of bad experiences, and critiques of the locals.[11] Human rights activists have not ceased to point to the lack of safety in work places, to the disastrous accommodation situation of migrant workers, and the often-illegal practices of employers such as depriving the worker of his or her passport.

In 2005, the government established the Qatar Financial Centre (QFC) and also the Qatar Financial Centre Authority (QFCA) by special laws. Following Dubai's free zone policy, the QFA should attract international financial services and other corporations. Meanwhile, Islamic banks have also settled there and benefitted from a rise in demand for *sharia* compliant products. The QFCA assures that the QFC works according to international standards and provides legal security. One hundred percent ownership by foreigners is allowed, and a low tax on profits of 10 percent is generated. (QFCA 2013) All these efforts are reflected in a good "Doing Business" rating. Qatar has reached rank 39 of the Doing Business rating in 2010 but droped to rank 50 in 2014. Major bottlenecks were getting credit, where it only ranked on 135 in 2010, and the enforcement of contracts. It is noteworthy that Qatar is on rank two (2010) for paying taxes. (World Bank 2010)

Also the institutional rating is quite sound and has constantly improved, as table 5 shows. Today, Qatar has on average received the best institutional ratings in the GCC states, and the corruption perception index assigns Qatar a ranking only slightly below Germany (7.7 compared to 7.9 for Germany).

***Table 5* Qatar's Governance Ratings**

Year	Voice & Accountability	Political Stability	Government Effectiveness	Regulatory Quality	Rule of Law	Control of Corruption
1996	-0.94	0.332	0.443	0.351	0.118	-0.24
1998	-0.75	0.965	0.673	0.209	0.487	0.797
2000	-0.54	1.033	0.575	0.105	0.705	0.84
2002	-0.62	0.651	0.615	0.302	0.741	0.844
2003	-0.67	0.951	0.605	0.178	0.55	0.768
2004	-0.47	0.915	0.717	0.19	0.605	0.812
2005	-0.4	0.875	0.466	0.271	0.751	0.875
2006	-0.72	0.923	0.479	0.298	0.745	0.888
2007	-0.73	0.897	0.134	0.461	0.752	1.015
2008	-0.69	1.051	0.85	0.772	0.895	1.328
2009	-0.89	1.118	1.13	0.624	0.959	1.639

Source: World Bank 2010.

One may thus ask why Qatar has not managed to reach higher non-oil growth rates. On average, its non-oil and non-gas sector has only grown by 0.29 percent on average. UNCTAD's concentration index for Qatar shows a steady but slow diversification of the market. From an average 0.631 in 1995, there were consistent improvements to 0.553 in 2008 and 0.522 in 2013.

Qatar's current development plan, Qatar Vision 2030 (General Secretariat for Development Planning 2008) may implicitly provide answers. As Hvidt put it, Qatar "is aiming for a future with oil and gas" (Hvidt 2013: 27) and Vision 2030 is marked by a tradeoff between economic diversification and a "reasonable" depletion of natural resources. Qatar can afford to plan its diversification slowly, and it puts a special focus on social justice and the preservation of traditions. The Qatari government can also afford to plan development with state-owned companies. The private sector should be enhanced, but rather in the distant future than a precipitately privatization of hitherto state-owned companies. (Hvidt 2013: 25-28)

Until recently, Qatar seemed to challenge the apparent unity of the GCC states by international moves that were not seen positively by its neighbors, most notably Saudi Arabia. Apart from its role as an international peace broker, Qatar has also hosted contentious Islamist preachers and shown sympathy for the Muslim Brotherhood. One of the most remarkable and successful developments is probably the establishment of al-Jazeera, a widely broadcast TV channel, notorious for discussing controversial issues without ever criticizing Qatar.

When income from oil and gas was low in the second half of the 1990s, Emir Hamad also prepared for elections of a new legislative body. A first cautious step was the municipal election in 1999 by universal suffrage. Every man or woman over twenty-five could run for office although none of the six female candidates were elected. (BBC 1999) The election of a national Consultative Council has not taken place yet. First, it was postponed to June 2010, then to June 2013, now elections are indefinitely postponed. Since oil and gas prices have risen, and with them Qatar's income, the democratization process has apparently come to a halt. In 2013, Hamad, aged 61, handed over power to his son Tamim bin Hamad al-Thani (*1980).

The Economy of Saudi Arabia

Saudi Arabia is the world's largest oil producer. The oil concession was signed in 1933 with Standard Oil of California (SOCAL) and benefitted Saudi Arabia more than other oil concessions in this region. British influence in the Gulf declined noticeably.

SOCAL founded the California Arab Standard Oil Company, which was to become the Arabian American Oil Company (ARAMCO) in 1944. Talented Saudi workers were trained further and even sent abroad for studies. However, no big educational initiative was launched yet. Also foreign Arab workers came to work in Saudi Arabia and this brought about a limited spread of Arab nationalist ideas, although they never gained a strong foothold. (Al-Rasheed 2002: 112)

In the absence of any other state bureaucracy or development plans other than oil, ARAMCO not only dealt with oil drilling, production and transport, but also with schools, hospitals and vital services. "ARAMCO filled a gap where public services, education and health facilities were underdeveloped and in some parts of the country virtually non-existent." (Al-Rasheed 2002: 96) The Americans lived there in towns built like typical American suburbs and had their own community life. Researchers sometimes argue that ARMACO became a state within the Saudi state. (Foley 2010: 211-257)

Saudi reserves are estimated at about 262.7 billion barrels. Most of the oil fields were discovered before 1970: Abqaiq (1940), Ghawār (1948/49), Safānīya (1951), Berri (1964), Zuluf (1965) and in 1975, Shayba. The biggest oilfield, Ghawar produces around 4.5 million barrels a day. From the beginning, it was clear that the Saudi reserves were immense and would become the backbone of the Saudi economy. The first refinery opened in 1941.

From 1972 to 1980, the Saudi government had the means to buy all ARAMCO shares, until the company was entirely state owned by 1980. Today, ARAMCO is the biggest oil company in the world.

At the time of the discoveries, people in Saudi Arabia lived from farming, agriculture and trade, while Ibn Saud had British subsidies at his disposal. Trade and important trading networks were mostly located in the Western provinces. Jeddah was the home of important merchant families, and the Hejaz, with the holy cities of Mecca and Medina, was economically more advanced and more cosmopolitan. (Niblock/Malik 2007: 48ff) Pilgrimage to the holy cities was for a long time the most important source of income for the Saudi government. Even today this revenue is not negligible, and Mecca has seen a widely-deplored construction boom. Sean Foley notes that the revenue from the pilgrimage made Saudi Arabia a rentier state even before income from oil. (Foley 2010: 23)

It is noteworthy that, more than elsewhere in the Gulf, there was a concern for the compatibility of Islam and development. King Faysal, who was the first Saudi king to make development a priority, had a vision of development that preserved "authentic" Islam: "Faysal's political conservatism was combined with a vision that Saudi Arabia can import technological expertise and modernise economically while remaining faithful to Islam." (Al-Rasheed 2002: 123) This concerns all parts of the development process: new technologies which had to be taken over from "infidels", the fact that many Western experts came to work in Saudi Arabia, education, new media and their effect on everyday life (television, radio) and the role of women. King Faysal may have claimed that it was the duty of Muslims to be at the forefront of international development. (Al-Rasheed 2002: 122f) However, the compatibility of Islam and modern society is still – maybe today more than ever – put into question by conservatives and Islamist forces. Hence, the various governments of Saudi Arabia have often shied away from reforms. The results are double-standard situations that can be found in the banking sector, in the judiciary, and in the educational system.

Serious development efforts started in the late 1950s. Several projects in education, transport and infrastructure were launched, but the bulk of planning started in the late 1960s, when the Saudi government adopted the practice of development planning by five-year plans.

In 1970, the Ministry of Economic Planning published the first five-year development plan. (Looney 1990; Niblock/Malik 2007) It consisted mainly of infrastructure projects and education, but it had been conceptualized when income from oil was still low. During this first five-year plan, oil revenues rocketed, changing the basis for the second development plan: health care and education were to become free for every citizen, social security measures were introduced, goods were subsidized, wages were raised, and interest-free loans were offered to private companies, agricultural enterprises, or just for private consumption. There was massive investment in infrastructure, but even then, there were bottlenecks at the ports and infrastructure had to improve faster. The second development plan was much more expensive than the first one, and 40 percent more expensive than planned.

In 1976, the Saudi Arabian Basic Industries Corporation (SABIC) was founded. It would turn the fishing village al-Jubail (al-Jubayl) into a new industrial city. Production began in 1981. SABIC comprises several companies that produce steel, polyolefin, polyethylene, polypropylene and other petrochemicals. (Auty 1990; Kubursi 1984; Seznec/Kirk 2010) SABIC's task is to coordinate the industrialization process in Saudi Arabia. To this end, it provides capital and equipment. As in other oil and capital-rich economies, it seems opportune to venture into energy-intensive industry projects, although Saudi Arabia does suffer from a relative shortage of natural gas. Apart from al-Jubail, Yanbu (Yanbu 'al-Baḥr) was basically built up from scratch. The small port developed into an important industrial city. It is one of the most important Red Sea ports in Saudi Arabia and the industrial city hosts other petrochemical companies as well as the Yanbu cement company. Both al-Jubail and Yanbu were designed for industrial production, and also accommodation and leisure opportunities for workers. Both host refineries (SAMREF and SASREF) originated from joint ventures. (Luciani 2012) These industrial cities are at the core of Saudi Arabia's diversification plans, and there is even a Royal Commission for Jubail and Yanbu. (RCJY 2013) It was hoped that private enterprises would further process the products of these industrial plants within Saudi Arabia, and indeed the number of factories in light manufacturing has risen. Compared to 850 factories in the industrial sector in 1981, 3,906 were licensed in 2006. (SAMIRAD 2013) Meanwhile, private Saudi investors have even invested in expensive projects in the oil and oil-downstream industry, until recently the exclusive domain of the public sector. A kind of national bourgeoisie has come into being, who in turn demand more investment opportunities within the Kingdom. However, there are still gaps in the mid-size business segments, and there is still a deplorable concentration of enterprises in the hands of a few wealthy families. (Luciani 2005) This concentration can become a hindrance for reforms. Already in 1994, Kiren Aziz Chaudhry found that economic reform and liberalization in the 1980s "was stalled almost instantly through opposition by business elites." (Chaudhry 1994: 1)

Another diversification effort, at least in the view of Saudi planners, was the expansion of agriculture. With respect to climate, soil, and rainfall, Saudi Arabia hardly offers good conditions for agriculture. From an environmental point of view, agriculture has devastating effects on the country's scarce water resources and huge amounts of fertilizers are used to boost the harvest. Nevertheless, Saudi Arabia made self-sufficiency in its food supplya priority for several reasons: there is the political fear of being dependent on food imports, and agriculture is considered diversification that also provides employment opportunities for Saudis. Therefore, the government subsidized this sector and allocated land at no cost to both individual farmers and big agricultural companies, the latter often run by foreign companies.[12] The drawback is that production costs are too high to be competitive: the government invested heavily in this sector by subsidizing the farmers and the Saudi Arabian Agricultural Bank provided interest-free loans. Furthermore, the government regulated the market for agricultural products, causing even more inefficiencies. It bought wheat at much higher prices, provided cheap electricity and water, and even granted import monopolies for some agricultural products. Thus Saudi Arabia bought food security at the price of a highly distorted market. Nevertheless, it succeeded in reaching a remarkable output. Agricultural production rose and Saudi Arabia has even become a wheat exporter. When Saudi Arabia cut back on subsidies in the 1990s, growth in the agricultural sectors slowed down drastically. (Niblock/Malik 2007: 121) However, the Kingdom can still cover big parts of its own demand of grain and fruit.

While the agricultural sector certainly employs the rural Saudi population, foreign labor is employed in the bigger agricultural companies. Seeds, labor and agricultural machinery have to be imported and it is doubtful if Saudi Arabia really has gained independence through the expansion of an unsustainable agricultural sector.

On the same grounds, industrial production in Saudi Arabia can be challenged. Apart from a putative comparative advantage in heavy industry (oil), there are also bottlenecks: apart from a gas shortage, the lack of highly-skilled and unskilled labor is problematic. Furthermore, there is a relative distance to markets where the products of its industry are needed. While the latter problem has somewhat changed with the rapid industrialization in Asia, Saudi products still face competition from other GCC countries and the international market.

Saudi Arabia's dependence on foreign expertise has remained great up to today, despite a constantly growing population. As in other Gulf countries, it seems that the nationals have not yet been able to acquire the necessary skills to replace foreign experts. Despite the Saudization effort to try to oblige companies to employ more Saudi citizens, it seems that there is widespread underemployment in Saudi Arabia, either overtly, covered by jobs in the many parastatals, or within the administration. (Hertog 2012) Official unemployment rates are at about 11 percent (while much higher for the female population) although unofficial estimates are much higher. (Hertog 2012) More than in other GCC states, the educational system is inappropriate to prepare students for professional life. While there are certainly some institutions of excellence, the bulk of schools are under the influence of the Wahhabi *ulema* who make sure that the curricula conform to their view of the Wahhabi doctrine. M. Prokop found that:

> [c]oncessions to conservative religious forces in the field of education, particularly in politically sensitive periods, forfeited the creation of a knowledge society and ultimately the future of the young generation. (Prokop 2005: 59f)

Even at universities – in humanities as well as in sciences – students have to attend religious classes. Attempts by the government to change and to "Westernize" the curricula met with fierce resistance by religious forces. (Prokop 2005)

Even if Saudis are skilled enough for a job, they may not be able to play a meaningful role in the often inefficient, highly centralized parastatals and administrative structures, where distance to power is large. (Hofstede 1980; Hamaiza 2015; Thompson 2015) However, even if demotivation is certainly a big issue, also the rentier mentality is at work. Data differs enormously and has to be treated with caution but it seems that only 34.1% of Saudis of working age are actually employed. According to Hertog (2012: 14), this is a "large-scale under-utilization of Saudi Arabia's national labour potential". This is partly due to the very young population and low female employment, but also to labour market segmentation and low educational resources. (Hertog 2012) Saudis are underrepresented in the private sector, and additionally employers prefer ex-pat workers. (Niblock/Malik 2007: 222f) The same holds for private sector activities. For several decades, the Saudi Arabian government has tried to encourage private sector activities, but still, the Saudi Arabian economy is mainly centrally planned. Private businessmen only seem to engage in business if they expect fast gains at no risk. (Looney 1992; id. 2004)

Until recently, plans to encourage Foreign Direct Investments (FDI) have not been very successful either.[13] There was FDI, but mainly related to the petroleum sector. The hoped-for diversification has not yet materialized. It has to be noted that FDI is not welcomed by the entire society. In the private sector, local businessmen fear the competition of foreign companies (Hertog 2005), and in society, resentment of Western influences has grown.

The Saudi capital market has gone through a long period of unclear regulations, doubtful brokerage practices, and potential losses due to the uncertainty of investors. Only in 2003 was a Saudi Capital Market Law passed and the creation of the Capital Market Authority put an end to a long situation of unclear competences. In 2001, the automated trading system *Tadawul* was established, and since, investments have been opened to foreign, non-Arab investors. (Ramady 2010: 155ff) As a result the Saudi stock market is today on the cutting edge of technology, and is considered mature and stable although suffering from some sectoral concentration. A similar development can be observed in the banking sector.

The banking sector has also been marked by a lot of caution and hesitancy. (Wilson, P. 1991) In 1976, there were only fourteen foreign banks operating in Saudi Arabia, compared to forty in Abu Dhabi and 50 in Dubai. Among other reasons for this, Rodney Wilson blames religious considerations:

> A second factor was the view that it would be unseemly for the country in which Islam's holiest places were sited to aspire to becoming a major international financial centre, with all that that implied concerning interest-rate policy and financial wheeling and dealing generally. (Wilson, R. 1982a: 282)

Later on, foreign banks were Saudi-ized, which meant that they became local banks with majority Saudi shareholdings.[14] In the aftermath, their scope of maneuver was extended. Despite the tacit ban on interest, *sharia* compliant products (based on commissions) were not successful at that time. Again, Wilson states:

> In Saudi Arabia the dilemma for the banks is that they have to comply with the spirit as well as the letter of Islamic law, and if commissions were changed too often, or raised too far, their role as a proxy for interest would be evident for all to see. (Wilson, R. 1982a: 294)

This hesitancy and also a lack of new sophisticated products in the Saudi banking sector led to an exodus of funds from Saudi Arabia to Bahrain, where banks offered much better opportunities. In general, the banking situation in Saudi Arabia was marked by a gap between the ideals of the law and of the practice.

In theory, there is a ban on interest in Saudi Arabia because all legislation has to be in accordance with *sharia*. (Looney 1990: 121-129) A comprehensive banking law did not exist, and only recently did the government tackle several banking issues by issuing royal decrees. (Ramady 2010) But this does not mean that there was Islamic banking from the beginning: as Rodney Wilson remarked, if there is no "unIslamic" banking there is no need to have special Islamic banks. All banks have to comply with Islamic law. (Wilson, R. 1982: 110; Ramady 2010: 127) In practice, the various banks have mostly offered interest-based products. The government acquiesced into this practice, but interest was not always unproblematic. During the economic crisis 1982-1986, many contractors who were not able to repay their (interest-based loans) appealed to the *sharia* courts where judges readily declared that the interest had been illegal. In the aftermath, many banks required government support to cover their losses. (Wilson, P./Graham 1994: 198f)

Banking disputes are often settled by the Saudi Arabian Monetary Agency (SAMA), which regulates the financial market rather efficiently. (Hertog 2005: 124; Ramady 2010: 79ff) The Saudi government succeeded in giving SAMA the independence to act according to economic necessities rather than political or religious considerations. For a long time, the Saudi financial sector was considered stable, but partly underdeveloped (Ramady 2010: 74) and not sophisticated enough to meet current demand. Some products have been barred due to religious consideration.

Mortgages were "illegal", and if they were offered, they were unsecured by law until the Saudi Mortgage Law was passed in 2012. For a long time, insurances were also unregulated, and only in 2003 an Insurance Companies Law came into force. (Hodgins/Beswetherick 2009) Although there are rapid improvements today, the insurance sector was for a long time underdeveloped and disputes regarding insurance apparently had to be settled by *sharia* courts. (IMF 2006: 24) Whether insurances comply with *sharia* is heavily debated. (Masum Billah 1993: 315f) An international law firm warned on its site that "[t]here is always the risk that a court or tribunal in Saudi Arabia may find that a particular insurance policy, particularly a bespoke policy for a large financing project, does not meet the requirements to be Shari'ah compliant." (Hodgins/Beswetherick 2009)

It is therefore not astonishing that a lot of Saudis place their money with Bahraini or Emirati banks. However, the fact that Saudi Arabia's banking sector opted by default for stability instead of venturing upon a huge portfolio of highly modern and sophisticated financial products may have been a huge advantage. The country's financial sector was affected by the financial crisis in 2008 but mostly only indirectly. (Ramady 2010: 118)

Currently, Islamic banking in Saudi Arabia is on the rise. Instead of establishing Islamic banks, existing banks have increasingly offered Islamic products. The preponderance of *murābaḥa* finance does not contribute to the esteem of Islamic banking in the eyes of Islamic scholars. This preponderance seems to be aggravated by Saudi Arabia's institutional weaknesses as well as by the obligatory alms tax. The alms tax is also levied on company profits, but companies widely evade the tax by declaring lower profits than actually made. Therefore, profit and risk-sharing products of Islamic finance are risky. (Kuran 2004: 13) It is unclear whether the obligatory *zakat* is to the advantage of the Saudi Arabian economy. The existence of a *religious* tax contributes to a picture of Saudi Arabia as one of the most conservative Islamic countries, more so than a normal income tax would.[15] Furthermore, some sources point to the possibility that *zakat* is used to finance militant Islamist organizations. (Prados/Blanchard 2004) What is more, *zakat* is often not collected properly because the *zakat* department lacks information on the population and their economic situation. Thus, *zakat* in Saudi Arabia does not even have the benefit of providing the government with statistical data for policies and research. (Arab News 2010)

Finally, the strict public morals in Saudi Arabia have also caused several losses for the Saudi Arabian economy. Apart from the pilgrimage, Saudi Arabia does not attract tourism. Non-Muslims are not allowed to enter the cities of Mecca and Medina and, without official invitation, foreigners are not allowed to enter Saudi Arabia at all. Getting a visa for Saudi Arabia is often a cumbersome, insecure process and a deterrent in itself. The UAE style luxury tourism is not an option for diversification.

Table 6 **Saudi Arabia's Governance Ratings**

Year	Voice & Accountability	Political Stability	Government Effectiveness	Regulatory Quality	Rule of Law	Control of Corruption
1996	-1.63	-0.55	-0.52	-0.34	0.251	-1.05
1998	-1.5	-0.11	-0.24	-0.18	0.18	-0.07
2000	-1.49	0.024	-0.21	-0.09	-0.01	0.133
2002	-1.64	-0.34	-0.31	-0.05	0.057	0.268
2003	-1.63	-0.19	-0.31	0.051	0.116	0.185
2004	-1.41	-0.88	-0.26	0.059	0.14	-0.12
2005	-1.32	-0.55	-0.36	0.099	0.077	-0.04
2006	-1.67	-0.54	-0.14	-0.07	0.083	-0.19
2007	-1.62	-0.49	-0.06	0.017	0.203	-0.1
2008	-1.68	-0.36	-0.01	0.126	0.212	0.17
2009	-1.77	-0.37	-0.09	0.223	0.123	0.146

Source: World Bank 2010.

The influence of Islamic conservatism on the legal institutions is considered paramount (Ballantyne 1980; 1986; 2006; Brown 1997), and this perception – no matter if it corresponds to the true situation or not – may be reflected in the low institutional ratings of Saudi Arabia (see table 6): compared to the other GCC states, the Kingdom stays consistently below the regional average. Despite low ratings of voice and accountability, looking at the potential "autonomy" of the government in economic decisions, government effectiveness and regulatory quality are rated low. Apparently, the Saudi government feels the urge to respond to demands of religious conservatives and the business bourgeoisie, which is far from being just a class of rent-seekers, (Luciani 2005: 151ff) but who nonetheless try to uphold their privileges.

Under King Abdullah there were been serious attempts to overcome these inefficiencies. In 2002, Saudi Arabia also gave itself a "vision" – long term development goals for the year 2020. Vision 2020 aims at economic diversification, an improvement of human resources, and a modernization of governance structures, while at the same time "preserving religious values and cultural heritage of the Kingdom." (MoEP 2004) In many cases these attempts to diversify were cautious, as the attempt to enhance tourism shows: in 2000, the "Saudi Commission for Tourism & Antiquities" was founded with an explicit goal to expand the tourism sector in order to diversify the economy and to create job opportunities. However, the website remains vague and it seems that only Saudi citizens are the target group of this project. The website states:

> The Saudi Council of Ministers issued Resolution No. (9), dated 12/1/1421 H (16/04/200 CE) to establish the Supreme Commission for Tourism (SCT), confirming the tourism sector as one of the productive sectors, and in order to contain the Saudi tourists within the country, increasing investment opportunities, developing human resources, and creating new job opportunities for Saudi nationals. (SCTA 2010)

In general, the Commission emphasizes the Islamic values of Saudi Arabia while avoiding any allusion to foreigners, the international public, or non-Muslims. Saudi Arabia actually used to have a tradition of tolerance towards foreigners and other religions, especially in the Hejaz. More recently, however, the doctrine of Islamists that only Muslims should live on the Arabian Peninsula has gained ground. (Foley 2010: 211-257) Also Lacroix (2014) argues that "Saudi society was largely secularized" in the 1950s and 1960s and that the role of religion in the state gained ground in the 1990s. Today, Saudi Arabia's government only takes very careful steps towards opening up the country. This cautiousness became awfully obvious during the twelve-year long accession negotiations, first to the GATT and then to the World Trade Organization (WTO). (Fürtig 2006; Hertog 2008) During the 1990s, WTO membership became increasingly important for Saudi Arabia as it faced budget deficits and the increasing unemployment of its rapidly growing population. Furthermore, all other GCC states had accessed the WTO earlier, either in the 1990s or in 2000 (Oman). What Saudi Arabia tried to negotiate, however, was the squaring of the circle: encouraging foreign direct investments, while at the same time protect its own economy in most sectors while minimizing the cultural influence and visibility of foreign influence. Apart from subsidies for local businesses as well as a long "negative list" of sectors foreigners should not be allowed to invest in, several political issues were problematic: the ban on companies with business relations with Israel and issues of terror financing needed clarification. Membership in the WTO therefore required several extremely unpopular steps. The private sector had to become competitive, without subsidies. Custom duties and taxes for foreign companies had to be lowered or abolished, and more sectors and branches were to open for foreign investments. In 2000, a new Foreign Capital Investment Law[16] was approved by the cabinet – just one law in a series of new business legislation. At the same time, the Saudi Arabian General Investment Authority (SAGIA) was founded as a one-stop shop for relevant business procedures. (Niblock/Malik 2007: 187f)

Eventually, Saudi Arabia even had to agree to deal with Israel according to WTO rules. The fact that the accession was finally successful in 2005 testifies to the determination of King Abdullah. Since then, several other reforms give room for hope: from June 2006 on, Saudi Arabia drastically simplified procedures to start a business, while other requirements were lowered or abolished. In addition, investor rights were improved. From rank 159 on the World Bank's Doing Business scale in 2007, Saudi Arabia climbed to 36 in 2008 and to rank 13 in 2010 but dropped again to rank 26 in 2014. Whoever has gone through a mere visa application process for Saudi Arabia or tried to get a bill paid may question the meaningfulness of these ratings and wonder how, within three years, Saudi Arabia overtook all other GCC states on the Doing Business scales. Furthermore, serious problems with contract enforcement remain, which could overshadow gains in other realms.[17]

Furthermore, King Abdullah aimed at reforming the judicial system. (Ramady 2013: 78) Official courts in Saudi Arabia are all *sharia* courts, even if most matters relating to business and banking are referred to special committees or even to SAMA for resolution. However, the responsibilities of these courts and committees are not strictly separated and there always remains a latent danger, even for foreigners, that a case will be heard by a *sharia* court should a Saudi party to the dispute demand such. The late Abdullah's reforms aimed at the creation of a real commercial court and an end of this uncertainty, stemming from the duality of legal systems. (Ansary 2008) In 2011, an Anti-Corruption Commission was established by Royal Decree to ease the obvious lack of accountability in Saudi bureaucracy. (Ramady 2013: 79)

My calculation of average non-oil growth for Saudi Arabia from 1973 to 2006 yielded a poor -1.09 percent which seems rather low given the tremendous efforts of industrialization. However, the concentration index also shows no improvement. From 0.686 in 1995, it reached 0.779 in 2008 and 0.767 in 2013. (UNCTAD 2014)

King Abdullah's moves towards diversification were not without opponents. For the liberal opposition the measures did not go far enough; for the conservatives many of them went too far. While the succession of King Salman, already reported to suffer from poor health, in January 2015 went smoothly, the stability of the country will not just depend on steps towards more participation. It will equally depend on the success of Saudization to create jobs for the fast growing population. Thus far success has been rather limited. (Hsu 2015)

The United Arab Emirates

In this chapter, I will focus on the two most important emirates of the UAE: Abu Dhabi (Abū Ẓaby) and Dubai (Dubayy). The other emirates, namely Sharjah (ash-Shariqah in Arabic), Ajman, (Ajmān), Umm al-Quwain (Umm al-Qīwayn), Ras al-Khaimah (Rā's al-Khaymah) and Fujairah (Fujayrah) have partially made efforts to diversify their economies (tourism) and have invested in education but the extent cannot be compared to either Abu Dhabi or Dubai yet. (Shihab 2001)

The Federation

When speaking of the UAE, it is noteworthy that this federation only came into being in 1971 and that, in spite of an unusually strong federal structure for the Arab World, economic policy is mostly decided at the level of the single emirates.

The competition between Abu Dhabi and Dubai goes back to the early 19th century, when the Maktoum family (today the rulers of Dubai) separated from the Al-

Nahyan ruled Bani Yas and settled in Dubai. There were times of fierce competition and even hostility between Dubai and Abu Dhabi, as well as times of mutual endorsement against a common enemy such as the Al-Saud or the Al-Qawasim in Sharjah and Ras al-Khaimah.

When Britain declared its retreat in 1968, credit was given to Sheikh Zayed bin Sultan Al-Nahyan, Amir of Abu Dhabi, for creating the federation of the UAE in 1971 (Ras al-Khaimah only joined in 1972). The fact that 92 percent of the UAE's oil reserves are located in Abu Dhabi and the willingness of Sheikh Zayed to share revenues with the other emirates turned the oil wealth into the "principe fédérateur" of the UAE. (Gibert/Maraut/Telle 2005: 19)

Politically, an asymmetric distribution of power is blatant. Only Abu Dhabi and Dubai have veto power in the Supreme Council of the Union and both hold eight seats in the Consultative Council although they considerably differ in size. This preponderance reflects the economic situation. The other five emirates have considerably less oil and gas and diversify very cautiously. Sharjah tries to keep pace with the "leading emirates": it invested in international education (American University of Sharjah), tourism, and the cultural sector. Despite Abu Dhabi and Dubai taking the lead, the emirates are aware that instability in one emirate can easily affect the credibility and reputation of the entire UAE. This is probably the reason why Abu Dhabi bailed out Dubai when its real estate market crushed in October 2008. Dubai's financial problems and its growing dependence on Abu Dhabi could even prove to be an advantage to the strength of the federation (Davidson 2009: 172-175), and recently the relationship between Dubai and Abu Dhabi seems to have been rather cordial. (Gibert/Maraut/Telle 2005: 59)

Nevertheless, there is reason for pessimism about the further integration and federalization of the UAE. While their interests coincide in many cases, there have been occasions when single emirates acted unilaterally on an international level. The attitude towards Israel highlights this problem. While Abu Dhabi, and accordingly the other emirates, adopted anti-Israel propaganda as elsewhere in the Arab world, Dubai established business ties with Israel that then became known within Dubai society.

Furthermore, it should be noted that new political institutions are usually those of the single emirates and not at a federal level. The judicial integration still leaves a lot to be desired with Ras al-Khaimah and Dubai having their own legal systems. (Al-Muhairi 1996) Additionally, all the free zones in Dubai were means to avoid federal legislation.

Economically, this heterogeneity is cumbersome. The Central Bank of the UAE, for example, has little possibility to survey money markets and banks in the emirates. Even if it detected problems, it could hardly act to counterbalance developments. Most emirates have had financial scandals and the Central Bank was unable to detect them in advance. Most alarmingly, many projects are doubled. Instead of benefiting from linkages among the emirates, it seems that every emirate launches its own projects. There are six international airports in the UAE that are not spread over the country, but are in the north-east of the country where six of the seven emirates are located. A similar development took place with internationally operating airlines.

The Economy of Abu Dhabi

As in the case of the other emirates, Abu Dhabi's history was marked by agriculture and pearling, but only to a lesser extent by trade. This may have been an advantage. When the British East Indian Company arrived, they had an interest in destroying the economic power of the Qawasim, whose trading activities in the Persian Gulf were a threat, while Abu Dhabi was not much of a competitor. Furthermore, it was time and again weakened by fratricide and power struggles amongst the Al-Nahyan. After they had crushed the Qawasim, the British controlled the economic development of Abu Dhabi. Great Britain precluded modernization by preventing new technologies from reaching Abu Dhabi. In pearling, Abu Dhabi fell behind other regions where pearl-diving already benefitted from modern technology. When Japanese artificial pearls entered the marked, Abu Dhabi's economy was hit hard. (Davidson 2009: 69-93) Further measures such as taxes and restrictions even led to an economic drain of Abu Dhabi, much to the benefit of Dubai, where some native Abu Dhabi-based merchants settled. As a result, Abu Dhabi's situation in the 1950s and 1960s was rather dim. The first modernization of infrastructure took place in al-Ayn (al-'Ayn) under Sheikh Zayed. With Shakhbut still ruling in Abu Dhabi, this created the rather unusual situation of a provincial town being more developed than the capital. (Davidson 2009: 30, 31 and 24)

The massive investments of revenue from oil into infrastructure started under Sheikh Zayed. (Ghanem 2001) This was facilitated by major oil and gas discoveries in the 1970s. From 1985 to 1986 the reserves rose from around 33 billion barrels to around 98 billion barrels, a number still valid today. Currently, Abu Dhabi produces around 2.7 million barrels a day, but it aims at raising its output to 3.5 million barrels from new investments. In 1971, the Abu Dhabi National Oil Company was founded. While there are several foreign shareholders (British Petroleum, Compagnie Française des Petroles, Royal Dutch Shell, Exxon-Mobil...), ADNOC has always had a majority. The same holds true for several ADNOC subsidiary companies that deal with transport and upstream operations, and also with downstream production (refinery, petrochemicals) and natural gas. Some of these companies are only of recent origin (BOROUGE, TAKREER). Though not comparable to the reserves of Qatar, Abu Dhabi also has considerable natural gas reserves. Altogether the oil and gas sector is still the crucial sector in Abu Dhabi's economy, and there are plans for expansion.

Like other Gulf States, Abu Dhabi has tried to make use of its comparative advantage in energy to industrialize. It produces metals (Emirates Aluminium, EMAL), plastics (Abu Dhabi Polymers Company), fertilizers (FERTIL) and other petrochemicals. A big part of the production is exported to Southeast Asia and this sector is constantly expanding: the Abu Dhabi Basic Industries Corporation is planning new industries (cement, iron) and several new plants.

Abu Dhabi's government has always given the greatest importance to overseas investments. It is estimated that Abu Dhabi's overseas investments total around one trillion US dollars. The Abu Dhabi Investment Authority (ADIA), founded in 1976, is still the world's largest sovereign wealth fund interested mainly in "save" long-term investments. Only a small part of the capital is invested in more risky ventures. Together with several other sovereign wealth funds, ADIA is formally controlled by the Abu Dhabi Investment Council. According to the Linaburg-Maduell Transparency Index, ADIA's transparency has improved from 3 to 5. Its current assets are estimated at US $627 billion. (SWFI 2014)

In 2002, Sheikh Muhammad bin Zayed al-Nahyan (today, crown prince) has founded the Mubadala Development Company. The company mainly invests in international technology companies such as the Carlyle Group, Advanced Micro Devices (AMD) and General Electric. It also holds a 5 percent stake in Ferrari, and recently Mubadala significantly diversified its portfolio. Its assets are estimated at $55.5 billion and its transparency is very high (10). (SWFI 2014) These are only the most important sovereign wealth funds. Indeed, many important figures in political life have their own consortia investing abroad.

Abu Dhabi's overseas investments are part of a crucial economic strategy and their significance should not be unterrated. Nevertheless, several problems pushed the rulers to consider local economic development more urgently. Employment opportunities for nationals had to be created, and the trade balanced having suffered from massive imports of consumption goods. While the last years under Sheikh Zayed were marked by stagnation, his successor Sheikh Khalifa tackled these problems more vigorously. Given the enormous oil wealth of the emirate, Abu Dhabi could opt for a cautious and smooth diversification process, without stirring too much social unrest. A "Plan Abu Dhabi 2030" has been set up, carefully identifying several long-term goals. With this plan, Abu Dhabi's government distances itself from other Gulf States and their short-term development plans, keeping the confusion that had resulted from the scarcely planned diversification in Dubai at the back of its mind. Abu Dhabi's "New Economy" is marked by some new features. First of all, instead of just investing abroad, Abu Dhabi has turned some of its parastatal companies into joint ventures with foreign high technology companies. The deal seems to be that Abu Dhabi provides the capital while the foreign companies offer the expertise. This strategy has proved to be very successful: non-oil exports rose by 50 percent in 2008. Prominent examples are deals with the European Aeronautic Defence and Space Company (EADS), General Electrics, and Rolls Royce. Not surprisingly, considerable stakes held by ADIA in international enterprises have led to some joint ventures, as in the case of AMD. Benefits are expected in the long run, as long as Abu Dhabi nationals grasp the opportunity to gain foreign expertise.

Another focus is on future energy industries. Abu Dhabi is among the worst per capita polluters, both in CO_2 per capita and in per capita waste. (Global Footprint Network 2010). Nevertheless, Abu Dhabi has expertise and experience in the energy business, and it aims to expand this to renewable energies. Partly funded by Mubadala, partly by joint ventures with foreign companies, the Abu Dhabi Future Energy Company (ADFEC) was founded in 2006, and its most prominent project is Masdar City. Masdar City is not only supposed to become a carbon neutral city, but also host to around 1,500 renewable energy companies. (Ouroussoff 2010) However, its finalization has been postponed to 2015–2020, and it might only become "low carbon" instead of "no carbon".

Together with international universities and companies, Abu Dhabi is preparing itself to become a storing place for captured carbon, using its disused oil wells as a storing space. This could become a viable business if carbon capture technologies become more widespread.

Abu Dhabi also discovered the real estate sector, but later than Dubai. While foreigners were not allowed to own property in Abu Dhabi until 2005, the government finally gave in and copied Dubai's renewable 99-year-lease contracts. However, Abu Dhabi's government made sure to avoid some of Dubai's mistakes: first, it made sure not to exaggerate construction in order to avoid overselling property, causing a potential price drop. Secondly, there is no desperate need to bring in as many foreign investors as possible, and Abu Dhabi can choose customers carefully. More than Dubai, it tries to preserve its traditional values, by segregating housing areas for nationals and foreigners. Non-Arab foreigners and a "corresponding" infrastructure of bars etc. will be located on Reem (Rīm) Island. Segregation also plays a role in the booming tourist sector. While the number of luxury hotels is constantly rising, planners try to make sure that nationals do not need to intermingle with foreigners. Abu Dhabi tries to serve the high end of the market and the best example may be the Emirates Palace hotel, which is hardly affordable for middle class foreigners. (Sharpley 2002)

In 2007, the Sheikh Zayed Mosque opened its doors – a huge and prestigious mosque located outside of Abu Dhabi. Today, this mosque is also accessible for non-Muslim tourists and has become one of the few cultural attractions in Abu Dhabi.

Another prestigious project is thought to compensate for a relative lack of cultural attractions in Abu Dhabi. The new island "Saadiyat" (Saʻadiyāt) is planned, to host a branch of the Guggenheim Museum and the Louvre, together with luxury hotels, shopping malls, and many other leisure possibilities. Combined with international sporting events and the invitation of international stars, Abu Dhabi hopes to compete with Dubai in the entertainment sector. These projects are to be complemented by the new Abu Dhabi-based Etihad Airline (founded in 2003), after Abu Dhabi withdrew unilaterally from its involvement in the loss-making Gulf Air. (Kassem/Habib 1989: 114-131)

A real problem for Abu Dhabi's economic development is its lack of skilled nationals. Few Emiratis have an academic degree, and young male students are especially underrepresented. Unemployment among nationals is also high and it is estimated that more than 50 percent of those Emiratis out of work are actually capable of working. (Davidson 2009: 149-153) Despite emiratisation policies, Emiratis account for less than 1 percent of the private sector employees and the young generation seems to continue to prefer public sector jobs. (Charles 2015)

Given its vast oil reserves, Abu Dhabi can afford this expensive ruling bargain, including the typical distributive measures: free education and health care, housing projects, subsidies for the poorer nationals and no taxes. These measures are completed by Islamic policies, alms giving, and the sponsorship of Islamic conferences and mosques. These policies seem to have been effective so far and Abu Dhabi's political system still works without meaningful popular participation. Abu Dhabi has reached a high degree of formal government, assigning tasks to several committees and institutions, but all of them are under close control of the ruling family.

The Economy of Dubai

Dubai's economy is probably the most puzzling case among the Gulf. Very early, by the end of the 19th century, Dubai became a free port that attracted many foreigners, mostly merchants from India, Pakistan and Persia. Abolishing taxes and restrictions, Dubai turned into a trade center and became the biggest city in the Emirates. The Al-Maktoum family was supported by Britain, but the fact that Dubai was very open caused problems, and soon the wealthy and important merchant class began to demand a say in politics.

In the first half of the 20th century, Dubai became a trading hub, specializing in re-exports. Merchants mostly traded in textiles and gold, and later on, electronics and cars. In some cases, important merchant families were granted the trade license for brand names (Mercedes cars, Sony), which they hold to today. Dubai benefitted a lot from trade restrictions in neighboring countries and it is hard to draw a line between the re-export business in Dubai and smuggling. (Davidson 2008: 67-71)

The immigration of merchant families from other emirates, India, Pakistan or Iran was another brick in the success story of Dubai: Dubai managed to attract and support the elites of other countries who already had experience in business who had fled trade restrictions. This emigration continues up to today. In the 1970s and 1980s, many Lebanese businessmen fled the civil war and went to Dubai. In 1990, Kuwaitis came to do business. Iranian emigration has basically never stopped.

Flourishing trade made it necessary to build infrastructure although Dubai had no income from oil at that time. The first mainly British-used airport in the region had been in Sharjah, but only few people were allowed to fly to Sharjah and then to go to Dubai. For a trading hub, this situation was unbearable, and in 1960 the Dubai airport was opened. Soon, it outweighed other airports in the region. Mostly the airplanes from Gulf Air – a Bahrain, Qatar, Oman, and Abu Dhabi-owned airline – landed in Dubai. In 1985, however, the governments of these countries criticized Dubai for its free air policy and demanded it equally pay subsidies to Gulf Air flights. Dubai refused, and as a consequence, fewer Gulf Air planes landed in Dubai. Sheikh Maktoum, however, took a bold step in a reaction to the pressure: He established "Emirates", a Dubai airline, which was soon to become more successful than Gulf Air, mostly by a strategy of offering luxury in economy class also, and by rapidly expanding its destinations. The founding of Emirates is somehow symptomatic of Dubai's diversification process. Dubai defended its liberal policy, sometimes taking bold steps forwards, without knowing if this was to become profitable. (Davidson 2008: 109ff)

Very early on Dubai established a framework for trade: since 1958, foreign experts elaborated upon the business laws, and international banks were brought in to the emirate. In 1966, the Dubai Petroleum Company started to produce oil from the offshore oilfields off the coast of Dubai. Dubai never had as much oil as Abu Dhabi and from the beginning it was clear that Dubai would have to pursue a real oil-independence strategy. (Davidson 2008: 100) As with other Gulf States, Dubai places assets abroad, but of course it could not do this to the same extent as Abu Dhabi. The Investment Corporation of Dubai was only founded in 2006 and has estimated assets of US $19.6 billion. Dubai Holding came into existence in 2004 and administers the largest part of the wealth of Sheikh Muhammad. Dubai International Capital is the international branch of Dubai Holding and was also established in 2004. Thus, international assets play a comparatively small role in Dubai's diversification strategy.[18]

As in other Gulf States, several light industries were developed. Dubai Aluminum (DUBAL) was founded in 1979, Dubai Cables (DUCAB) in 1981, and a natural gas (DUGAS) plant was also established. More than other Gulf Countries, Dubai also tried to substitute for imports. Goods needed for construction and daily life (plastic bottles) were produced. Agriculture was also boosted, though not as much as in Abu Dhabi.

To diversify, Dubai concentrated mostly on the service sector. A remarkable step was the creation of free zones. In 1985, Dubai founded its first free zone "Jebel Ali". UAE legislation stipulated that a company must be owned by a national by at least 51 percent. For foreign investors this meant that they needed a *kafīl*, a local sponsor who received a salary just for this representative task. Because of the clear will to attract foreign capital and companies, Dubai circumvented this law by creating zones where it claimed that federal laws did not hold because they were free and independent and therefore not a real part of Dubai. While this decision was certainly not on firm ground, the guarantees given by Sheikh Rashid were enough to attract many international firms. Soon, other free zones followed: there is an Internet City for IT companies, a media city where internationally important media have branches and also the Dubai Knowledge Village, where international universities offer programs. The financial free zone, Dubai International Financial Center, has attracted many foreign banks and provides a British-inspired legal environment. While most oil-rich countries place their assets abroad, Dubai hopes that the foreign expertise hired will attract investors to use the Dubai International Financial Center to place assets there. It is therefore important that Islamic financial products are also offered by Dubai based banks. According to estimates, one third of Dubai's territory is made up of free zones today. (Davidson 2008: 114-118)

Dubai has also become a popular tourist destination, although tourists can hardly expect a rich cultural heritage. Rather, Dubai attracts tourists to its large shopping malls, horse and camel races, international sports events, and its mild winters. The credo seems to be to offer the superlative: the biggest, best and most beautiful of everything. It is thus not astonishing that most hotels are in the higher categories and that the luxury hotel Burj al-Arab has become a kind of national symbol of Dubai meanwhile. Other mega-projects such as the construction of the two artificial palm-shaped islands and the luxury hotel "Atlantis - The Palm" help to attract international attention to the region, even if they may be unprofitable. So far, the number of tourists has constantly risen. (Davidson 2008: 119-128) These projects are mostly run by the state-owned real estate company Nakheel (Palms). Nakheel, in turn, is a subsidiary of Dubai World, which is also an important international and local investment vehicle of the Dubai government, aiming at boosting Dubai as a trading hub. But Dubai World fell deeply into debt, creating an international uproar, and could only cover its debts with a US $10 billion loan from Abu Dhabi. (BBC 2009)

The famous Burj Khalifa (formerly Burj Dubai) was built by Emaar Properties. Rumor has it that it was named Burj Khalifa, because Abu Dhabi's Sheikh Khalifa's financial help was unexpectedly necessary to finish the tower.

The real estate sector in Dubai deserves special attention. The construction sector in Dubai has not ceased for several decades now. Apart from offices, Dubai also reckons that foreigners will also buy apartments and houses. Companies might buy apartments for their employees, but also millionaires and wealthy people from less stable regions might have an interest in a refuge. Again, Dubai circumvented federal legislation stipulating that only nationals can own real estate in the UAE. Formally, this problem was solved by a leasing contract of 99 years, which can be renewed by the symbolic payment of one dollar. The most important factor seems to be the confidence in the ruling family of Dubai, who guaranteed the right to buy property to foreigners. (Davidson 2008: 99) For a long time demand exceeded supply and big profits could be made by speculation: many houses, apartments and offices were only bought to be re-sold, a highly profitable business providing owners with rents. (Davidson 2008:128-132) The real estate bubble burst in 2008 when the international financial crisis finally reached Dubai. Real estate companies stopped projects and their shares dropped in the country's stock market. It seems that a good deal of the diversification achieved by real estate was just "paper sand", driven by speculators. However, optimists argue that 2008 has brought a healthy readjustment of prices which had quadrupled over the preceding five years. And despite this sudden end to the boom, it seems that Dubai has established itself too firmly to be abandoned.

So far, Dubai's economic development seems to be a success story. Christopher Davidson sums up the development process quite cautiously:

> Taken as a whole, these diversification strategies would seem to have greatly reduced Dubai's reliance on oil exports and oil-related strategies such as energy dependent heavy industries and oil-financed overseas investments. By the mid-1990s Dubai's non-oil sectors were already contributing 82 per cent of the

> emirate's GDP, and in 2006 the figure may have been as high as 95 per cent. [...] Thus, on paper at least, Dubai would appear to have escaped from its former dependency and certainly gives the impression that it has succeeded in creating a vast wealth, independently of the old oil industry, which is now most often described as 'simply ticking over'. (Davidson 2008: 135)

Nonetheless, Dubai's economy is highly dependent on the world economy. This may be true of any country in the world, but for Dubai, this dependency is exacerbated.

First of all, Dubai has not escaped the pathologies of the rentier state. A big part of the national population seems to live from "post-oil" rents, for example, from renting out real estate or from securing jobs in the government administration or in parastatal companies. Few of these jobs are demanding and the youth does not seem to aspire to more. In 2002, Sheikh Muhammad initiated a widespread check on nationals at their working place and realized that many did not appear on time or did not appear at all. As a consequence, many employees, also on the senior level, were sacked. Apparently, the privilege to work little while earning a lot has engrained itself also in the mentality of Dubai nationals. This might be the reason why attempts to employ more Emiratis in the private sector have not been successful yet. Foreign employers prefer less demanding, higher skilled and hard-working foreigners to Dubai nationals, who moreover are hard to release once hired. Thus, the dependency on foreign expertise will probably not end in the near future. A Dubai based merchant even states:

> ... of Dubai's young nationals, probably only 20 per cent will be worthwhile, becoming academics and professionals, and businessmen. About 60 per cent can probably be written off, the consequences of the all-too-easy acceptance of the pleasures which will be handed out to them. (Davidson 2008: 179)

Among these "pleasures", over-consumption should be mentioned. Whoever spends time in Dubai realizes that the most important national occupation is shopping, even if this is often hidden behind a cultural pretext. The local demand is so high that even despite the import substitution efforts, Dubai and the entire UAE import much more than they export when oil and gas exports are not taken into account. When oil and gas prices are modest or low, this deficit remains even after adding fossil fuel exports.[19]

A second problem could result from the fact that Dubai's young nationals have never experienced an extractive state and would probably oppose any measures of taxation. Dubai's oil reserves are estimated to last for another twenty years only, and probably the government will need other sources of income then. Up to today, Dubai has taken several cautious steps, such as tiny indirect taxes and taxes affecting mostly foreigners.

The diversification may have reduced the dependency on oil, but it has increased the dependence on foreigners, and especially on wealthy foreigners. The tourist sector depends on foreigners who can afford luxury, but they might fail to appear in times of economic depression.

Also the ideological bargain between the rulers and the national population, and hence political stability, is increasingly at stake. There are nationals who have come to resent the strong presence of migrant workers and their disrespect of Islamic holidays, and who frown on the widespread consumption of alcohol in night-clubs. The weekend has been changed from Thursdays and Fridays to Fridays and Saturdays – a clear assimilation to the Western weekend, which was nevertheless common in many foreign companies. This situation may also be a fertile ground for Islamism and it is not unlikely that terrorist groups might use the liberal climate of Dubai to organize their activities. There have been several reports on funds for terrorists being channeled through Dubai. What is more, it cannot be ruled out that Dubai – as other rulers in the Gulf probably do – pays "fees" to terrorist groups to make them refrain from terrorist attacks in the Gulf.

Additionally, organized crime seems to thrive in Dubai. Money-laundering, drug-trafficking and human-trafficking are reported as widespread in Dubai with no effective government counter-measures being initiated. (Davidson 2008: 282-292)

Despite these problems, the power of the Al-Maktoum seems to be secure. The family is closely-knit and there seem to be no significant cleavages. The transition from Sheikh Rashid, who died in 1990, to his son Sheikh Maktoum, was smooth despite the lack of a fixed succession regulation. The same was the case after the unexpected death of Sheikh Maktoum in 2006, when his brother Sheikh Muhammad succeeded him and was elected vice-president and prime-minister of the UAE.

The UAE in International Ratings

As the short overviews of the economies have demonstrated, there are differences between Abu Dhabi and Dubai and even more so between Abu Dhabi and Dubai and the other emirates. Investment guides and business oriented agencies (e.g. Standard's & Poor's) take this fact into account and provide reports on the single emirates or single institutions. By contrast, World Bank ratings refer to the entire UAE. The World Bank governance ratings assert the UAE a rather strong institutional environment and political stability, on the same level as Qatar for example (see table 7).

The Doing Business rating for the UAE is rather mixed. Apart from good ratings for taxes, across-border trade, or registering property, there are strangley low ratings for investor protection, contract enforcement and closing a business. (WB Doing Business 2010) It should be noted, however, that these ratings refer to heterogeneous situations of the single emirates and to different legal environments resulting from the free zones. This heterogeneity makes planning and forecasting difficult on the federal level. Meanwhile, the UAE emulated the examples of other GCC states, publishing its own "Vision 2021" (in 2021, the federationwill celebrate its 50th anniversary), but this "Vision" remains vague. (UAE Cabinet 2010; Hvidt 2013: 31f)

Table 7 **The UAE's Governance Ratings**

Year	Voice & Accountability	Political Stability	Government Effectiveness	Regulatory Quality	Rule of Law	Control of Corruption
1996	-1.02	0.736	0.384	0.54	0.64	-0.25
1998	-0.56	0.731	0.831	0.594	0.826	0.387
2000	-0.58	0.798	0.787	0.692	0.802	0.528
2002	-0.72	0.755	0.854	1.079	0.78	1.135
2003	-0.82	0.838	0.811	0.82	0.707	1.159
2004	-0.78	0.693	0.806	0.85	0.664	1.175
2005	-0.68	0.688	0.56	0.495	0.523	1.03
2006	-1.01	0.852	0.778	0.685	0.423	0.94
2007	-1	0.903	0.918	0.68	0.377	1.033
2008	-0.95	0.709	0.877	0.641	0.545	1.082
2009	-0.87	0.907	0.927	0.557	0.525	1.037

Source: World Bank 2010.

The document is certainly inspired by the "Dubai Model" and the strategy to reach high growth rates in non-oil sectors, but it also asserts a lot of flexibility in reaching diversification. On the whole it is not a groundbreaking vision, stipulating real changes in the economic structure. (Hvidt 2013: 32f)

Although the development trajectories of the seven emirates are clearly very heterogenous, older data is usually available for the UAE only. I calculated an average non-oil growth rate from 1973 to 2006 of 7.09 percent. This seems to be too good to be true, but even at 3 percent the performance would have been very high, and the overall judgment of the rapid diversification process remains. This is also reflected in the improvement of the concentration index: In 1995, the UAE were assigned a 0.563. In several years, it dropped below 0.4 and even to 0.331 in 2009. In 2013, it was at 0.423.

2.3 The Gulf Cooperation Council

Before the creation of the Gulf Cooperation Council in 1981 there was an informal intergovernmental cooperation between the Arab Gulf states (often including Iraq). Economic agendas were at least partly harmonized and there were several common economic projects. Examples are the Gulf Ports Union (for the improvement of the ports), the United Arab Shipping Company (founded 1976-77), the Gulf Petrochemical Industries Company (1979), the Gulf Aluminium Rolling Mills Company (1981), and the joint purchase of what was to become Gulf Air by Bahrain, Qatar, Abu Dhabi and Oman in 1973. However, this initial cooperation was soon replaced by duplications: while ALBA, the Bahraini aluminium smelter, was a joint venture with the Bahraini government holding 77.9 percent of the shares, other GCC states built up their own aluminium production. Only very recently plans to merge aluminium industries were announced. (Ramady 2013: 258f) The same duplication occurred with Gulf Air; all GCC states now have their own airlines. The ASRY shipyard, Bahrain's ship repair yard, was another project all Arab Gulf states (save for Oman) invested in. Nevertheless, Dubai opened its own ship repair yard complex in 1979 – in direct competition to ASRY. (Townsend 1980)

The regional threats of the late 1970s and early 1980s (Iranian Revolution and Iraq-Iran war) gave rise to the creation of more formal cooperation. (Kechichian 1985) It was Saudi Arabia that put its weight behind such a project and, in 1981 the Gulf Cooperation Council was established for this purpose. Iraq was not invited to join, a clear sign that this would be a club of conservative monarchies, and it seems that the GCC will remain such a "club". Recently, two other Arab states, Jordan and Morocco, made a bid for membership. (Toumi 2011)

Institutionally the GCC is governed by the Supreme Council, and is ruled by the heads of the member states. The Council has the decisional power in any matter, with every member having equal voting power and the right to veto decisions. The Ministerial Council consists of the foreign ministers of the member states. There are also other committees, made up of the member's ministers. The Committee for Financial and Economic Cooperation is a regular meeting of the GCC's ministers of finance and economy; similar committees include the ministers of industry, agriculture, environment, development and planning gather in their respective committees.

The dominance by Saudi Arabia cannot be denied: the Secretary General's office is located in Riyadh, and Saudis hold the most important positions. In May 2009, the GCC members agreed that the precursor of the GCC's central bank, the GCC Monetary Council, would be located in Riyadh as well. To be sure, this triggered objections in the UAE, probably because Abu Dhabi had been a strong candidate. (Hamdan/Pantin 2009)

Aims were formulated for all possible fields of cooperation: more economic cooperation, further liberalization (especially among the GCC states), freedom of residence of GCC nationals in member countries, trade cooperation, common customs, and eventually, a common market and single currency. The Unified Economic Agreement of 1981 aimed high when declaring all these goals, but no clear steps were specified at this stage.

It is noteworthy that interest in economic integration in the Persian Gulf differs across the GCC. The six Gulf States all subsidise their industries and many of their industrial projects would probably have ceased to exist if it had not been for the subsidies. (Koppers 1995: 52) However, since the GCC states are unequally rich, some states can subsidise their industries more than others. This of course can create problems if markets are liberalized, especially because these economies are thus far not complementary, but competitive. Importantly, potential benefits from economic integration could be dwarfed by the public sector. Publicly owned companies or parastatals are less likely to respond to incentives presented by economic integration.

This is not the place to dwell on theories of economic or political integration, although I would like to point out that two different "directions" mark these theories. Some argue that integration can only take place once several commonalities have been reached; others claim that first a cornerstone has to be laid to trigger a harmonization of policies. In the case of the GCC states, opinions also differ with regard to the potential benefits from economic integration: Whereas some argue that there would be no benefits because the economies all produce the same goods and all suffer from the same diseases, others claim that integration and the resulting competition would force the GCC economies to develop niches to find their own comparative advantages. (Koppers 1995: 73-77)

Discussions of the potential benefits remained theoretical because economic integration keeps being postponed. During the twenty years since its inception, the GCC remained at the lowest degree of integration – the Free Trade Area – and even this level was not completely achieved. In 1983, customs were abolished among the GCC states, with the exception of Oman. Customs tariffs for products from third countries were partly harmonized at a low level: they had to range from 4 to 20 percent, but were not harmonized. In addition, every country had a vast array of exceptions: products "necessary for daily life" were not subject to any tariffs, while protective tariffs were imposed on products likely to compete with those locally produced. (Koppers 1995: 179-184) No customs union was thus reached during this time. Additionally, all GCC citizens were granted the right of domicile in all GCC states. While at first they were only entitled to buy property for private use, they later received the right to buy property for business purposes. However, there were several restrictions hampering this "right". While economic rights were expanded step by step, specifications on a national level often made it difficult to benefit from these agreements. As for badly needed economic cooperation due to the increased duplication of projects, little progress was made. Maybe one of the most noticeable achievements of the GCC was a widespread acknowledgement of the GCC as a distinct "Arabian" region as opposed to an "Arab" or "Middle Eastern" frame of reference. (Legrenzi 2008)

Conducting a large survey of experts as to achievements made by the GCC to 1990, Simon Koppers interviewed forty economists from the region. The findings point to gaps in the implementation of the few above-mentioned achievements. Respondents did not think that the Free Trade area was really implemented, and they still noticed major obstacles to free trade. The same holds true for the freedom of economic activities within the GCC. In practice, there were discriminatory laws making it extremely unattractive for GCC nationals to do business in a GCC state other than his home country. (Koppers 1995: 128-133) There is thus reason to believe that the few apparently binding stipulations of the GCC agreements were circumvented and it is doubtful if any positive effects have resulted from such cooperation. (Koppers 1995: 156)

Sefik Bahadir (1990) draws a disappointing picture of the 1980s. For him, cooperation before 1981 suited the habits of the region much better than within an institution such as the GCC, which was not given any supranational capacity. On the international level, however, the GCC states have negotiated as a group on several occasions although very few of these negotiations were actually successful. GCC countries united when buying from international firms only in rare instances. The different airlines, for example, all bought their planes separately when they would have been in a much stronger bargaining position if they had negotiated jointly. Neither were endeavors to reach free trade agreements with important sales markets crowned by success. The European Economic Community, for example, signed a Cooperation Agreement[20] with the GCC, but it stipulated that a Customs Union should be achieved before a free trade agreement could be negotiated. A customs union, however, was only implemented in 2003. While negotiations were relaunched in 2002, no EU-GCC free trade agreement has yet been reached. (European Union External Action 2014) As for GCC-EU relations in general, Koch found that a "quantitative improvement has [...] not translated in to a qualitative one as well." (Koch 2014:1)

Although customs unions take their time, even more so for common markets, the many announcements that were not followed by action call for explanations. The reluctance to give up absolute sovereignty is paramount. This reluctance is understandable, especially when the efforts the GCC governments make in order to co-opt their societies are taken into account. They are obviously not eager to expand the rent distribution to nationals of poorer neighbors. Furthermore, they shy away from confronting the old merchant families that have not just trade but industrialism interests also. As such, they favor protective tariffs as well as complicated or barred market entries for potential competitors.

Eventually there would be winners and losers in an integration process. In the long run, the less resource-rich states face the risk of being unable to subsidize their economies sufficiently enough to be competitive. It would have been an option to establish a fund to redeem the "losers" for their losses due to integration, but it seems that the GCC states prefer to grant support unilaterally and without any obligation. (Koppers 1995:205-216)

Recently, when oil prices rose again, cooperation re-intensified. (Sturm/Siegfried 2005) In 2001, on the occasion of the Muscat summit, a new Economic Agreement was adopted with the concrete aim of a common market and a single currency within the GCC. *De jure*, all GCC currencies had to be pegged to the US $ by 2003. *De facto*, this had already been the case: only the Kuwaiti dinar fluctuated within margins of 3.5 percent around the central rate. As for fiscal convergence, fluctuating incomes from oil makes it difficult to design appropriate fiscal criteria. Nevertheless, convergence criteria for the monetary union were specified in 2005, largely following the model of the euro. After the customs union had been put into practice in 2003, the "common market" came into effect in January 2008. The single currency – most likely the "*khaleej*", after the Arabic word for "Gulf" (*khalīj*) – was announced for production by the year 2010. By 2009, however, many analysts pointed to institutional weaknesses and continued nonconverging policies, and it came as no surprise that the monetary union was postponed. (Schubert 2010) A monetary union can only be viable if there is a strong political commitment of all members involved, a GCC central bank with supranational capacities, and a clear, independent mandate. So far, only the Gulf Monetary Council in Riyadh has been created, and it remains to be seen if the plan to create a common currency for the GCC by 2015 will be fulfilled. (Ramady 2013: 76)

Bahrain, Kuwait, Oman, Qatar, Saudi Arabia, and the UAE are also members of other regional and international organizations having similar objectives, but none of these organizations has reached the level of integration of the GCC – as low as those may be.

3

The GCC States and Economic Theory

The preceding chapter has highlighted some of the specificities of the region. The following chapter will show to what extent the GCC economies thereby challenge common theories of economic growth and development. The latter have resulted in a huge amount of research on the potential drivers of economic growth and development. As a result, the list of factors suspected of driving or impeding growth has grown massively over time.[1]

For the sake of a parsimonious model, it is important to keep in mind two things: firstly, the focus on oil exporting states provides some restrictions. The theories I will adduce have to be viewed in light of their context and their underlying assumptions even if they claim to be universal. Secondly, a theoretical background for an *empirical analysis* which aims at investigating the *conditions* of non-depletable growth is needed. I am concerned with a comparative view on the statics and not so much with the dynamics of the subject. I will not dwell, therefore, on reproducing here formal growth models, their assumptions and their derivations *at length*. (Bretschger 1999; Barro/Sala-i-Martin 2004; Frenkel/Hemmer 1999; Zhang 2005)

The following theories will be used to select variables for an empirical model that draws on *parts* of several theories since no single theory has given a detailed and comprehensive explanation of non-depletable growth in oil-exporting countries so far. Many aspects of these theories will not be taken into account, either because they are not relevant for resource abundant countries, or because they cannot be operationalized in my empirical model.

In this chapter, I move from the more general theories of economic growth and development economics to theories focusing on resource-abundant countries. The following two subsections deal with growth theory and development theory. These topics cannot be easily separated, and to add confusion, both have had important implications for development policy. Many authors seem to interchange these notions. For this analysis I propose a very simple but still quite arbitrary differentiation: theories will be referred to as growth theories when they have originated from a context of developed, industrially advanced countries, even if they deal with these countries' "undevelopment". Theories that have been formulated with a special focus on less developed countries (LDCs) will be consequently referred to as economic development theories. They deal with "underdevelopment" in comparison with other, developed countries.

3.1 Growth Theory and the GCC Economies

Economic growth can be defined as "the increase in a country's standard of living over time." (Howitt/Weil 2008) This normally goes hand in hand with an increase in value of the goods and services produced by an economy, and an increased output per worker. Growth theories deal with factors that drive growth and try to explain growth rates under different assumptions.

The origins of growth theory are usually traced back to Adam Smith (1723-1790) who described how productivity of labor grows through a division of labor with the result that wages rise. Increasing wages leads to more demand and, therefore, to an expanding market. In return, more demand makes it profitable to produce more. If production exceeds demand, prices will decrease, reducing now unprofitable production. This is the famous "invisible hand", regulating the market automatically on the condition that there is a good government that guarantees rule of law and an absence of monopolies. Although Smith's view was far more complex, divison of labor and the principle of *laissez-faire*, the non-intervention of the government, henceforth became the basic ingredients of classical and neoclassical economics.

Thomas Malthus is known for his pessimistic view on population growth. Malthus observed that the amount of land (capital) remains stable while populations grow. As a result, more and more workers will work the same amount of land. This increases productivity at first, but when labor supply increases further, productivity will fall and workers will earn less. In this case, diminishing returns of labor would finally result in a situation where people work again on a subsistence level. The idea of holding one factor of production stable while adding the other factor of production that in turn leads to a lower increase in productivity than the unit before, is called the "law of diminishing marginal productivity". Apart from stopping population growth, Malthus saw only one way out: technological change in agriculture, which to him seemed to be a rare event.

Among the classics, also David Ricardo's work is worth mentioning here. In Ricardo's model, there are capitalists, workers, and landlords. Capitalists invest and accumulate capital as long as their profits are positive. Workers do not possess any means of production and, according to Ricardo, their amount changes with their wages. Higher wages lead to rising birth rates and decreasing death rates, which result eventually in a larger work force. More workers, in turn, result in lower wages and lower population growth. Wages will finally reach a "natural rate", which corresponds to the subsistence level.

The landlord class charges rents for the use of their soil. For Ricardo, rents will be as high as the difference between the output of a hired, fertile unit of land and a poor, freely available and equally big unit of land given a fixed amount of labor and capital. In today's rentier states literature, this so-called Ricardian rent is hardly mentioned any more.

For Ricardo, the capitalist is the driving force of growth. Thanks to capital accumulation, output and wages will grow. Population will grow rapidly, and even if this stalls because of lower wages, the population will expand for a certain time. Consequently, more and more land will be worked, and rents will decrease. There is less profit to make for the capitalist and eventually, capitalists will only be able to pay wages without making profit. Thus, the growth process comes to a halt and a stationary state is reached. Like Malthus, Ricardo saw the possibility of improved techniques in agriculture but assumed that these improvements would only slow down the process. (Baldwin 1966: 23-25)

The classics have provided the very basics of modern orthodox economics. Developing an economy based on the division of labor instead of the nomadic subsistence economy of the Bedouins is certainly an important step. In the GCC countries, however, there is no Smithean *laissez-faire* state. The GCC governments pursue a form of capitalism that is called "state-sponsored". (Niblock/Malik 2007: 7-10) Instead of precluding monopolies, the governments sometimes even create monopolies, and makes use of market-distorting subsidies.

Population growth, as in Malthus's theory, is actually considered a threat for the GCC economies, but the mechanism behind this assumed negative influence on economic growth differs significantly from the mechanism assumed by Malthus. In today's GCC economies, technological change can be imported and is freely available, as is capital. Technological change has taken place at a dizzying pace and the productivity per worker has skyrocketed compared to the productivity of nomads and Bedouins. If population growth is problematic for the GCC economies, the reasons are entirely different from those assumed by Malthus. Firstly, a slowdown of productivity does not necessarily lead to worse living conditions anymore. The extent of the welfare state in the GCC countries makes sure that living conditions and economic growth (not based on higher oil prices or increased production) are almost uncoupled. If there were a cutback on welfare, this would instead be due to lower oil prices or the depletion of a country's reserves. And secondly, population growth does not necessarily mean that the labor force increases: people may lack the necessary skills or they may not need to work because their living standard is high enough without having to work. Modern technologies require not only a quantitative rise of the labor force, but also a qualitative improvement.

In the first half of the 20th century, several economists independently developed influential growth models. In 1911, the Austrian economist Joseph Schumpeter published his *Theory of Economic Development*. Schumpeter proposed a cyclical growth model where the driving force is the entrepreneur. The entrepreneur is the driver of innovation and technological change when he or she puts his/her ideas into practice, mostly by new combinations of factors of production. To do so, the entrepreneur not only needs what Schumpeter calls "Unternehmergeist" (entrepreneurial spirit), but also a well-developed financial system to provide the seed capital. (Audretsch; Keilback; Lehmann 2006) Schumpeter considered large corporations as change resistant. To apply untried technologies, therefore, establishing a new company was necessary. What was needed was "creative destruction" – old ways of production had to be overcome, and substituted by innovative methods. (Schumpeter 1942) Schumpeter was rather pessimistic about the future of entrepreneurship. Successful entrepreneurship would increase prosperity but thus also the advent of new elites with moralistic views. This, in return, would create an environment hostile towards entrepreneurship, and eventually, to its decline.

The era of Schumpeter was also marked by John Maynard Keynes (1883-1946) who can hardly be called a growth theorist, but whose reflections on unemployment and uncertainty have implications for growth. (Lepenies/Nitsch 2000) In his *Treatise on Money* (1930) Keynes had already rejected the neoclassical assumption of savings equalizing investments. Saved money could also remain "uninvested" and therefore non-productive.[2] The reason for the hesitancy of investors was their uncertainty about the future.[3] Less money invested meant, for Keynes and many other economists, lower growth, or even recession. While Keynes at first thought that the gap between savings and investment would have temporary effects only, the Great Depression taught him that this was too optimistic an opinion. Rather, economic declines lead to lower wages and unemployment. Lower wages, in turn, lead to smaller margins for savings and lower investments. In such a situation, firms have no incentives to invest because – and this is Keynes' "uncertainty" – they do not expect higher sales. When an economy reaches this situation, Keynes suggested that the government could overcome the stalemate by creating demand as an exceptional short-term measure when necessary, even through deficit spending. Public demand (e.g., public infrastructure orders) would convince investors that sales are on the rise. According to Keynes, it is effective demand which creates supply. (Riese 1983; Singer 1999)

Even if Keynes deviated manifestly from usual assumptions of neoclassical economists, his macroeconomic theory was the starting point for what became known as the neoclassical growth model. While the latter is usually attributed to economists of the post-war period (Robert Solow, Trevor Swan) the foundations were laid by the post-Keynesians Roy Harrod (1939) and Evsey Domar (1946). Both independently reached the same formal model, describing the natural growth rate of an economy in its steady state (an economy where demand equals supply, also in the labor market) under perfect competition. The key element of the Harrod-Domar[4] model was investment, which equalled, in a deviation from Keynes, savings. This simple model became very popular, and with it, capital and investment became the one-and-only driving force of growth. The lack of both would mean poverty without the means of change. Proponents of development aid argued (and still do) that poor countries could not attain sufficient growth rates because they did not have enough savings. (Easterly 2001: 28ff) Harrod and Domar did not exactly create a growth model. They gave a formal description of how growth can take place when several knife-edge conditions were fulfilled and several parameters held constant. They did not explain growth, however.

Population growth (increase of labor supply) and the savings rate were exogenously given. In addition, the model may have explained short-term business cycles, but for the long run, the assumptions of a constant capital coefficient and a constant saving rate are not realistic. (Bretschger 1999: 24-25)

Even if hard to operationalize, Schumpeter's innovative entrepreneur – or rather the lack thereof – is of great interest for the GCC states. "Dynamic, innovative, risk-bearing, Schumpeter's entrepreneur is the antithesis of the rentier." (Beblawi 1987: 86) The main entrepreneur in the GCC states, however, has been the state. Governments hoped that, after an initial stage of public investments, private entrepreneurs would further the investment process. Recently, several state-sponsored programs have tried to lure young locals into entrepreneurship. This, of course, presupposes that there are profits to be made and that these profits will be higher than for other activities. However, in a distributive state people often gain more by rent-seeking. In the GCC states, businessmen could amass fortunes by proforma activities, such as sponsorship. Also extremely lucrative were import monopolies. These factors make true innovative and risky entrepreneurship less interesting.

Furthermore, Schumpeter failed to foresee one aggravating factor. The GCC states lagged behind technologically: rather than inventing and innovating themselves, they import new technologies from elsewhere, and the incentives to "innovate" are rather low. It even seems that the GCC states are trapped in an ongoing need to always have to catch up on foreign expertise and knowledge, which may distract them from their genuine innovative capacities.

It is also noteworthy that the GCC economies provide a new view on Keynes' deficit spending. Budget deficits have occurred and there was deficit spending, but in general GCC governments can invest from a pool of rents that have not been saved before. The rents stem from capital depreciation and not from savings. Again, resource abundance distorts assumptions that may be valid elsewhere. This insight is reflected in recent attempts to correct saving rates for capital depreciation and negative externalities, as in the concept of "net adjusted savings". (World Bank 2014)

The neoclassical growth model is mostly attributed to Robert M. Solow who laid its foundations in his seminal paper "A Contribution to the Theory of Economic Growth" in 1956. Solow claimed that, in the long run, an economy could not grow just by capital accumulation and population growth. At a given technology level and labor force, economic growth would cease because of diminishing returns to capital (Solow 1956)[5] until the economy reaches its steady state. Solow's model explains growth until this steady state is reached. Once the steady state is reached, growth can only be attained by technological change. Extending the Harrod-Domar model, Solow added labor as a production factor as well as technological change as a variable, affecting the production function as a whole.[6]

As for the factors driving growth, Solow also recognized labor and capital. His additional contribution was technological change, which, however, was exogenous and "fell like manna from heaven". It has to be imagined as an upward shift of the production function.

This crucial point – plus the fact that the other factor influencing growth, population growth, is equally exogenous – gave the model the name "exogenous growth model". (Bretschger 1999: 39; Frenkel/Hemmer 1999: 56f) To sum up, growth in the Solow model can only occur if a country has not reached its steady state, or if technological change takes place. (Frenkel/Hemmer 1999: 164)

As an implication of the Solow model, poor countries are considered to be far from their steady state level, and should grow faster than countries close to their steady state. This gave rise to the so-called convergence debate. Convergence means that poorer countries, growing faster, should eventually catch up with rich countries. Empirically, a tendency towards convergence could be found within the group of OECD countries. On a worldwide level, however, the contrary is the case. Very poor countries tend to stay poor. Capital did not flow as was expected from rich to poor countries, where labor is cheap, but tended to flow to emerging markets only. These results directed the attention towards other factors influencing growth, such as free markets and education. The result was "conditional convergence": only if all parameters of the model are the same in all countries, can the catch-up effect until convergence occur. (Bretschger 1999: 47; Frenkel/Hemmer 1999: 139-155)

Growth theory after 1945 did not yield new drivers of growth. Rather, the modeling was further elaborated. Again, the overall importance of capital is emphasized, which, in the GCC states, evokes the "paradox of plenty". The fact that resource-abundant states have the means to invest in capital formation but do not necessarily perform economically well, is at the heart of an ample body of literature on the resource curse, Dutch disease or rentier state theory, which is hardly in line with common neoclassical growth theory.

Technological change in an industrialized country is quite different from a country that, fifty years ago, had neither electricity nor hot water. This is the interesting point about the exogeneity of technological change. In today's developed countries, technological change had to be invented. Furthermore, it could take decades until a certain problem could be resolved or old methods improved. By contrast, in developing countries technological change does not result from local needs, research, or learning by doing: It is imported and therefore indeed *exogenous*. Coping with the new leaves little space for their own innovations. While considered to be a big flaw in Solow's model, the exogeneity of technological change can be quite accurate in developing countries. It externally arises, sometimes even against the will of the population. The corresponding social transformations may therefore differ considerably from those that took place during industrial and technological revolutions in Western countries.

Growth theory experienced another revival in the 1980s. The availability of the Heston and Summers data set (Heston/Summers/Aten 2009) facilitated empirical studies and revived the convergence debate. Many economists were dissatisfied with the exogenously given, and hence unexplained, technological change in Solow's model. These and other shortcomings led to the development of the "endogenous growth theory". The main objective of this theory is to explain long-run growth by endogenous variables. Growth should be generated from within the system. The starting point is the basic assumption of diminishing marginal returns of capital. Growth in the long run would be possible if marginal returns of capital were not diminishing but constant. What can prevent diminishing returns? (Bretschger 1999: 80ff)

Key concepts of endogenous growth theory are human capital, investments in research and development (R&D), and knowledge spillovers. These concepts are far from new. It seems obvious that labor is not only the amount of workers and the hours they spend working as they also have differences in knowledge and experience, but it took some time until human capital was included in growth theory. Investments into human capital can take the form of education, and also of R&D. R&D aims at developing new and better production techniques and increases public knowledge. The investment in R&D has an even bigger effect when the results benefit other sectors or firms by positive spillovers. Knowledge is different from physical capital in that it is a non-rival good and can be used by many people or firms at the same time. However, if everyone used new techniques at the same time (non-excludability) there are little expected profits, incentives to invest in research are low, and this market failure usually requires governmental action. Consequently, endogenous growth theory has to allow for at least partial monopolies in order to create incentives to invest in R&D. (Bretschger 1999: 91) In this case public services are needed, such as research programs, investments in infrastructure, legal security or even patent protection.

Moreover, the production process itself can give rise to more efficient methods. Workers gain more experience, make fewer mistakes, and find out how they work most efficiently. Knowledge, therefore, is "input and output of the production process". (Bretschger 1999: 85) Hence, the more capital accumulation and production, the more knowledge capital can be accumulated. (Arrow 1962)

Human capital, knowledge, and positive spillovers are the bricks and mortar of the endogenous growth theory. They are supposed to hold returns of capital constant, or increase them. Several models are based on endogenous technological change; others use spillovers to hold returns to capital constant. (Frenkel/Hemmer 1999, 173-180; Aghion 1998) The most prominent pioneers for endogenous growth are Paul Romer (Romer 1986; id. 1990; id. 1994)[7] and Nobel laureate Robert Lucas. (Lucas 1988) For this study, however, it is not so important how the models work exactly, but how technological change, positive spillovers, and human capital can be measured. (Wößmann 2004)

Human capital is a crucial point for the GCC states. A little more than half a century ago, few children went to elementary and even fewer to secondary schools. Universities did not exist. Meanwhile, GCC governments have realized the urgent need for more education. School enrollment and literacy rates skyrocketed in recent decades with Gulf citizens supported by government scholarships when they enrolled in tertiary education abroad. While the technical skills and the business know-how were desired souvenirs from the West, cultural adaptation was not.

Today, major western universities have satellite branches in the GCC states, their programs tailored to the "real needs" of Gulf societies. (Lewin 2008; Willoughby 2008) They are supposed to produce engineers, doctors, and computer scientists. For Gulf citizens, again, there are no tuition fees. One might expect that the demand for skilled workers could soon be met by Gulf Arabs, as the various localization policies stipulate. (Baghat 1999: 132) The failure and circumvention of these policies, however, give rise to concerns. G. Baghat describes the problems of education in the Gulf monarchies as follows:

> [T]he curriculum of schools and colleges continue to be dominated by Islamic and Arabic studies (Abir: 35). A large majority of students opt for the less demanding humanities and social science courses where the teaching and textbooks are mostly in Arabic. Second, at all levels of education there is much more emphasis put on academic learning than on vocational and technical training. The latter has been unpopular among Gulf youth and attracts only a small proportion of students. [...] Moreover, this unpopularity of vocational training can be explained by the public's attitude. Many people in the Gulf still value white-collar jobs much higher than the blue-collar ones. A new perception needs to develop. (Baghat 1999: 130-131)

Garrett and Farghaly (1987: 321) diagnose an "Arab aversion to manual and technical labor". They hardly play an innovative or at least active role in the private sector. Accordingly, positive spillovers and learning-by-doing benefits fail to appear.

What is the problem? Maybe one root of the problem has already been considered: technological change can be imported, and there is no need for the locals to become agents of innovation in order to improve their living conditions. Furthermore, not only can technological change be imported, but also the experts and workers. GCC governments hoped that their nationals would acquire the necessary skills step-by-step and be able to work with new technologies autonomously, or even to further develop technologies. However, not only is the innovative work left to foreign experts, the productive work is also mostly left to foreigners. Still, learning by doing among the locals is limited as are positive spillovers due to a daily experience with the work process.

Educational policy suffers from the lack of incentives and a weakness of coordination. While spending on education is quite high, graduates are not replacing expatriate labor. I will have to come back to these phenomena when human capital is operationalized.

3.2 Theories of Economic Development and the GCC Economies

Theories of economic development can be divided into two categories. The first category – endogenous theories – concentrates on the economic factors in one country and does not pay much attention to international interactions or history. The second category contains exogenous theories that concentrate on influences of a country's past and its position in the world economy. They claim that exterior factors are responsible for development problems. As a third approach, institutional economics is, strictly speaking, not a theory but an approach directing the views of the researcher on several political, economic and social phenomena that enhance or impede growth.

The neoclassical recipe for economic development is trade and openness for trade, i.e., trade liberalization. The roots of this almost dogmatic view lay with the theory of comparative advantage, which is usually attributed to David Ricardo (1962: 130ff) but was later further elaborated by the Swedish economists Bertil Ohlin (1933) and Eli Heckscher. (Maneschi 1998) For development economics, the main implication of the Heckscher-Ohlin-Theorem is trade specification. Comparative advantage means that every country is endowed with a specific set of factors of production. Some countries, for example, dispose of huge amounts of raw materials while others are resource poor. According to the comparative advantage theory, the country with the lowest opportunity cost of production should produce the product. All countries could thus benefit from mutual trade on the condition that they specialize in the production and export of the right goods. Therefore, international trade must be free, and LCDs have to open their economies.[8]

This is a very elegant theory but it relies on several questionable assumptions. While there must not be any barriers for the goods, capital and other factors of production (labor) are not supposed to be internationally mobile in this model. (Daly 1999: 32) In an international interconnected world, this is not conceivable. (Borchert 1975: 141) Furthermore, the model neglects transportation costs.

In the context of industrial development in the Gulf States, the comparative advantage argument is of crucial importance because of the oil and gas reserves that are – in the main – easy to recover. From the viewpoint of comparative advantage, they had better remain oil and gas producers and exporters while importing other goods from countries with other comparative advantages. Shortages of other factors of production (e.g., skilled engineers for resource production techniques) put this comparative advantage into question, and it cannot be denied that there are also other valuable objectives such as self-reliance in food production and a diversified "portfolio" of produced goods in case one market is affected by shocks. (Beutel 2012: 34ff) Furthermore, international factors of production *are* mobile. This is more than evident from labor immigration to the Gulf countries.

Jaime Ros (2001: 2) highlights that the "classical development economists", as he calls them, pointed to three important conditions that deviate from neoclassical assumptions: imperfect competition, increasing returns of capital, and labor surplus. Imperfect competition is due to market failures, which are often even aggravated by governments in developing countries. The neoclassical assumption that information is free and immediately available to anyone has been criticized by numerous development economists. Moreover, there is often a labor surplus due to high unemployment rates and a huge amount of people working in the unproductive subsistence sector. Increasing returns-to-scale occur in economies with very low capital intensity. Adding another unit of capital still has increasing effects on the output. The point when constant, and later diminishing, returns occur is not yet reached. In addition, while neoclassical growth theory was increasingly concerned with the steady state, classical development economists deal with quite a different kind of "steady state": poverty trap – a low-level equilibrium that, for many reasons, is also hard to overcome.

This problem has also been noticed by the Austrian economist Paul Rosenstein-Rodan.[9] As Rosenstein-Rodan saw it, people's decisions to invest depend on their expectations of investments by other people. As long as a potential investor does not expect others to invest, he or she will hesitate too. Thus, investment is a process that can easily develop a negative dynamic and reach a low-level equilibrium trap. Rosenstein-Rodan saw the solution in a detailed centrally-organized development plan. Investment had to flow into several sectors in a "balanced" symmetric way as a kind of "big push". The negative equilibrium of no investments because of no *expected* investments could, according to Rosenstein-Rodan, be "pushed" above a point where positive dynamics could develop. If the government invests, private investors have the security that investment "happens" and their money can grow on fertile ground. Furthermore, the first investments by the government would create jobs and a higher income, thus stimulating demand. The Big Push-Theory was criticized from various sides. Liberals frowned at the idea of government planning. Development specialists doubted that young governments and administrations were capable of such a widespread coordination effort. A famous critique came from Albert O. Hirschman, who claimed that such a long-term development could not be planned entirely in advance. Only the development process itself would show where the next investment was required. Hirschman developed his own concept of "unbalanced growth" relying on "linkages". Based on an already existing "leading" sector, forward linkages (downstream), backward linkages (upstream, capital production) and fiscal linkages (government income from taxes to be invested in new industries) should be created. Hirschman's thoughts stress the importance of the special situation, both in the sense of the right point in time and in the sense of each country's special and sometimes "hidden" resources – the "built-in modifiers or remedies" which had to be found and activated. (Hirschman 1958; 1995)

Nevertheless, Hirschman also proposed a "big push", but only from one leading sector. Secondary sectors should be developed by private investors in response to the needs of the leading sector. These are striking parallels with the development projects of the Gulf countries. Two of them (Saudi Arabia and Oman) have followed consecutive planning in clusters of five years and all of them have recently adopted long-term strategic "visions". GCC governments tried to initiate an investment chain, starting with huge public investments into their (leading) oil and gas sector. Contrary to Rosenstein-Rodan's balanced growth, they did not invest as much into other sectors, hoping that this would finally be taken over by private investors according to needs, and to the new demand created by the leading sector. However, the high benefits from rents make these investments appear less lucrative. Despite these problems, the GCC states are among the few states that could really afford to follow such a development path. According to Krommer, who analyzed the GCC development strategies in the late 1980s, development by investment into a leading sector characterized the entire GCC development strategies. (Krommer 1986)

Quite to the contrary, a fixed development path of five stages was proposed in 1960 by Walt Whitman Rostow. *The Stages of Economic Growth: A Non-Communist Manifesto* (1960)[10] was certainly one of the most influential publications of development economics. Based on historical analyses, the stages describe how a traditional society evolves into a society of "mass-consumption". The crucial step of a consumption-society is the third one: the famous "take-off", when barriers to modernity are removed and "growth becomes its normal condition" (Rostow 1960: 36). The last phase is marked by an increasing demand for luxury goods since the basic needs have been met. (Preston 1996: 175-177)

Rostow's five stages fail to reflect the situation in the GCC states and they are probably the best known example of the questionable assumption that all countries have to undergo the same process. The GCC states have reached a stage of industrial development without the revolution of society and Western modernity. Interestingly, however, the description of Rostow's last stage – the age of mass-consumption – may come very close to the reality in the GCC.

Modernization theory and other endogenous development theories have been criticized for their narrow viewpoint. A country does not develop in a vacuum, but as part of the international economy. It is surprising to find the absence of the international setting in many development and growth theories, even more so as writers pointed to this fact as early as at the beginning of the 19th century.

Friedrich List, for example, anticipated many development issues. Preconditions of development were free entrepreneurship, rule of law, and education, as well as national governments and administrations taking the right decisions and being capable of implementing them. (Senghaas 1999: 164-168) It seemed natural to him that development could only be the result of an advantageous constellation of factors and decisions adapted to the special situation of a country. Unfortunately, List stood for only one policy advice retained in posterity: protective tariffs. List wrote in the early 19th century, when Britain was about to industrialize, and Germany and the rest of Europe lagged behind. The danger seemed obvious: cheap British merchandise would glut the German market before Germany even had the opportunity to thrive. This is how the "infant industry argument" was born. However, List propagated protective tariffs only for special industries and for short periods of time. The competition of foreign goods was also a necessary incentive to industrialize, and the import of foreign capital goods did convey information about new production techniques.

The writings of Karl Marx were used to endorse *several* development policies. Most importantly, he argued that imperialism was the result of fierce competition between capitalists. Their need of cheaper production factors, such as raw materials and labor, made them not only exploit workers in Europe, but increasingly workers and resources abroad.

Because capitalists needed only primary products to produce the refined goods themselves the colonized economies were destroyed and kept at a primary product level. The imperialist powers had no interest in further economic development of the colonies. This idea inspired a very influential school that evolved after the Second World War. Indeed, the colonial powers frequently used their territories overseas as suppliers of primary products and as sales markets for their industrial goods. Britain, for example, imported cotton from Egypt and then sold cloth and clothes cheaper than the local producers. The latter, facing competition from industrial production, could not survive. In the long run, therefore, only the primary sector had a chance.

This kind of unequal trade relationship sometimes survived colonialism. South American states gained their independence at the beginning of the 19th century but continued to export primary products (beef and wheat) and even experienced a remarkable economic upswing. Industrial goods were imported, however. It was the elite of European origin who especially benefitted from this liberal economic policy. During the Great Depression prices in general fell, and even more for primary products. Up to this time the Argentinian economist Raúl Prebisch had advocated free trade, but experiencing the Great Depression in 1930, he came to support protectionism and Keynesianism. Later on, as director of the Economic Commission for Latin America, he formulated his thoughts as policies.

In his structuralist approach to the development problem of Latin America, Prebisch preferred a realistic model at the cost of formality and universality. For him, the world was not a uniform marketplace but divided into a powerful center with a weak periphery. The countries of the periphery were marked by heterogeneous economic structures due to the legacies of colonialism and a conflict of interest between sectors. Unlike modernization theory, he claimed that poor countries were not at an earlier stage of a fixed development path (that of the industrialized countries). They were structurally different due to their insertion in a changed world economy. Preston describes this as follows:

> The distinctively national and integrated economy does not exist and instead there is a collection of residues, enclaves and various parasitic forms. In the light of these analysis it can be argued that an 'underdeveloped country is underdeveloped precisely because it consists of different structures each with a specific type of behaviour.' (Preston 1996: 184)

In addition, Prebisch criticized the fact that the center could choose where to buy the required primary products. The consequence is the daily uncertainty of the peripheral countries about selling or not selling their commodities.

Later on, Prebisch added the financial sector to his theory. The major financial institutions were controlled by capitalist countries, and furthermore, all of the Bretton-Woods institutions were located in metropolitan cities in the USA. Also, he discerned a financial center and periphery.

The idea of a center-peripheral relationship was the foundation of what came to be known as the Singer-Prebisch Thesis. Prebisch and Singer observed independently that the price of primary products declined in relation to the price for manufactured goods. For Prebisch the reason was technical progress that added more and more value to the raw material, giving more importance to labor and less to the material at origin. This means that peripheral countries, bound to primary exports by the structure of the world economy, faced deteriorating terms of trade. Every time the price for their primary products declined, they could afford fewer manufactured goods.

Prebisch drew several policy conclusions, centering on the industrialization of the periphery. Step by step the manufacturing sector was supposed to supplant the primary product sector. Manufactured goods which had hitherto been imported would be produced locally. Thereafter, the technical advantage of the center would diminish. (Preston 1996: 180-186) In Latin America, this strategy of import-substituting industrialization was indeed successful at the beginning. Later on, several problems, such as dependence on foreign supply, small local sales markets as well as over-enthusiasm, corruption and petrodollars, led to a debt crisis in the late 1970s. (Preston 1996: 187ff)

Prebisch's thoughts were crucial for the development of dependency theory (Thirlwall 2003: 133-142; Palma 1978), which is an umbrella term for several currents. Leading "*dependistas*" were Fernando Henrique Cardoso, Osvaldo Sunkel, Theotonio Dos Santos and André Gunder Frank, who is sometimes considered the "father" of dependency theory. Today, Dieter Senghaas and Kunibert Raffer are advocates of dependency theory, and argue that the "North-South Divide" still prevails. (Senghaas 1974; Raffer/Singer 2001)

André Gunder Frank is probably the most radical of the *dependistas.* (Preston 1996: 218f; Hein 2000) In his early works, he criticized modernization theory because of its assumption that there was only one path of development. (Frank 1971) There is a fundamental difference between underdeveloped countries in the 20th century and non-developed countries more than a century ago. Today's developed countries have never been *under*developed. They may have been *un*developed, and, this is Frank's radical position, they had developed at the expense of the peripheral countries, which then *became* underdeveloped. (Frank 1969)

A common point of critique states that dependency theory focuses only on external reasons for underdevelopment, i.e., the integration of the periphery into the world economy, when the center dictated the terms of trade. For Frank, however, a part of the problem was domestic: the domestic elites, in turn often the result of colonial rule, were the domestic link in a chain of exploitation. Dependency had been internalized and was no longer a purely external matter. (Raffer/Singer 2001: 35f)

It is unclear whether it was really colonialism that precluded the development of economic sectors other than the primary products sectors in the GCC states. Their inclusion in the capitalist world economy took place very late and the main interest of British influence was, at first, not the economy, but security. This changed only during the first half of the twentieth century, when oil was discovered. At that point, the exploitation of oil became interesting – not only for British and American oil companies, but also for the local rulers who wanted to gain from the concessions. Without foreign expertise, these rulers would not have been able to exploit their enormous wealth under the soil or sea. Hence, without being exploited by a colonial power, the GCC governments would not have had the money to invest into diversification and development later. Within a few decades they had acquired enough wealth to nationalize their oil companies. The wealth gave them a certain degree of independence from the "center" and at times even some power over the center.

The difference may lay with the kind of commodity: oil and gas are strategic commodities and their supply limited – at least in the past – to a few countries. Furthermore, the price of oil and gas has risen in real terms, even if price drops occurred at times. Thus, the exporters of the primary oil and gas products do not necessarily become poorer. Through their numerous sovereign wealth funds, they have become shareholders of the foreign companies who, in the view of dependency theory, are supposed to exploit them. The GCC not only imports consumer goods such as Mercedes cars, they are also large shareholders of Daimler AG, and benefit from the company performance and profits. At least for large oil exporters, the situation is more complicated than assumed by dependency theory.

Nevertheless, the GCC states still import most final goods, while their own products are rarely destined for the end consumer. There is very little import substitution, and the GCC states are a region with a high demand for foreign (luxury) goods and also enough purchasing power. Is this due to "colonial influence", or are resource abundant states a special case, calling for another theoretical explanation? Jean-François Seznec argues that the Gulf societies are basically trading societies: merchant families gain from trade and their trade monopolies more than from local production. Competing with highly sophisticated goods produced in the West would have been difficult at the beginning, and this dependence on foreign goods remains. (Seznec 2010: 2f) There is an international supply and interest in the GCC as a sales market, and there is a strong local interest, influential within royal and *sheikhly* circles, who want this situation to continue.

Although there is little consumer goods production in the GCC, the governments have protected their industries with subsidies and protective tariffs that distort industrial production. These measures have hardly been part of a comprehensive strategy to make local goods competitive. At the beginning of economic planning, these subsidies were supposed to make industrial production and entrepreneurship in manufacturing more attractive for the national population. Today, these subsidies can be considered as a part of the privileges for nationals, but not part of a comprehensive import substitution strategy.

Institutionalism was originally an American intellectual movement of the late 19th and early 20th century that deplored universalistic pretensions of economics as held by neoclassical approaches or by moralist American apologists.

Institutionalists held the opinion that economic behavior and patterns were influenced by history and the changing institutions of society. Institutions can adopt various forms.

There are rules of society that differ in sources, scope and enforcement. Simple conventions, social norms or ethical commitments nobody is *obliged* to follow are institutions as well as formal laws that are enforced if necessary. What is important is the fact that institutions *guide* the decisions of individuals. An individual's set of choices is reduced either by merely a way of thinking, moral feelings, the fear of bad reputation, or even punishment by law.[11]

Institutional economics has greatly influenced development economics. Gunnar Myrdal (1957; 1970), the outstanding voice of institutional development theory,[12] should also be mentioned in this context. Myrdal opposed the view that every country had to follow the same path of development and that international free trade would sooner or later level out differences between developed and underdeveloped countries. To those who favored growth theories and the "invisible hand of the market" he objected with two reasonings. For Myrdal, any economic system was embedded in a society. The institutions of society influenced development as much as the availability of capital. Any change was a complex process with various interplays. It seemed necessary to him that economic analysis be open to sociology, political science, anthropology and history. Additionally, Myrdal asked why poor countries seemed to become poorer and rich countries even richer. To him, the equilibrium models of growth theory, where a trend into one direction would immediately provoke the opposite trend, were not realistic. Positive factors in a society were likely to have positive consequences whereas negative factors provoked negative effects. If a society experiences growth, people become richer and can afford better education. Better education, in turn, enhances the chances for more growth and more wealth. This is an example for what Myrdal calls "circular cumulative causation". Again, these are poverty traps: widening gaps between the rich and the poor, both within a poor country as well as internationally. In order to overcome downward sloping economic processes, the entire social system of a country has to shift towards a positive trend. This is Myrdal's understanding of development, which, according to him, requires state ordered planning. (Preston 1996: 200-203)

In 1975, the notion of "new institutionalist economics" (NIE) was coined by Oliver Williamson. (Klein 2000; Voigt 2003; Clague 1997 and Clague/Keefer 1997a) Other well-known representatives are Ronald H. Coase (Coase 1984) and Douglas North (North 1989; id. 1990; id. 1991). New institutionalist economists share a lot with early American Iinstitutionalism, especially the request that economists have to broaden their view and include institutions in their analysis.[13] Both new and old institutionalists requested, in principle, an inductive way of reasoning, and rejected the neoclassical application of ideal equilibrium models to the reality. (Klein 2000) By contrast, new institutionalist economists do not shy away from applying neoclassical methods to their broader view on economics and society. The attempt to empirically compare real cases and to use the results in order to improve neoclassical theories is widespread. (Berg 2003; Flörkemeier 2003)

The NIE focus on individual decisions, and make two main assumptions: bounded rationality and positive transaction costs. While neoclassical economists usually assume the "*homo oeconomicus*" who takes his decisions following a profit-maximizing strategy that can be rationally understood, bounded rationality describes the fact that many decisions are not perfectly rational, but partly emotional. (Nutzinger/Panther 2003; Voigt 2003)

The second assumption is the existence of positive transaction costs. Transaction costs can be described as fees which have to be paid in exchange for the use of the market. They arise every time a good is sold, bought or rented. It is costly to gain information, to contact potential partners or clients, and negotiate transactions. Transport is costly and so are quality checks or cancellations. Neoclassical orthodoxy neglects these costs. In the ideal neoclassical market, there is no uncertainty. Every price has its quantity of buyers, which will rise as the price declines or decrease as the price rises. In the NIE market, however, there is uncertainty and therefore, there are transaction costs, which can have a negative effect on economic performance. The higher they are the fewer transactions take place. Transactions costs are usually very high if only few transactions take place and if the legal background or practical procedure is unclear. This can be either a pioneer problem, which means that after some experience transaction costs will decrease; or, it can be an ongoing problem of instability or big distances – be they local or cultural. This is where institutions can be helpful. "Institutions reduce uncertainty by providing a structure to everyday life," writes North. (1990: 3) Property rights are a good example. If they are guaranteed, transactions can take place freely and contracts can in most cases be enforced. This security makes transaction costs much lower and encourages people to buy, sell, and cooperate. Cultural commonalities and institutions raising trust, for example, reduce transaction costs because negotiations tend to be easier and there are fewer sources for (costly) misunderstandings. In a society marked by mistrust and uncertainty, people will cooperate less for fear of being betrayed. (Annen 2003) Division of labor and economies of scale (requiring big firm sizes) can only be the result of cooperation, and institutions that guarantee the enforcement of contracts if necessary.

This is where development economics can tie in. According to the new institutionalist economists, economic development needs trustworthy institutions in order to reduce costly uncertainty. In LDCs, there is a general lack of trust in governments and bureaucracies. There are no incentives to invest if the property of the investor is not secure or if every bureaucratic step is difficult or costly. Corruption distorts the efficient allocation of resources and deters potential investors or entrepreneurs. This is all the more the case for potential *foreign* investors. But corruption is both the reason for and the result of weak institutions, and this vicious circle is difficult to interrupt. (Lambsdorf/Taube/Schramm 2005)

In addition, harsh competition among actors in the market leads to unfair behavior, and this raises mistrust. Without cooperation, higher developed products, requiring technical know-how and a long array of production steps, will not be produced. Uncertainty and mistrust because of a lack of trustworthy institutions are, according to NIE, the main reasons for the ongoing prevalence of primary products in developing countries. (Clague 1997a: 27)

This is also supported by data. Clague, for instance, has made the empirical finding that:

> poor countries that get their institutions and policies right will grow faster than rich countries, because the poor ones have greater opportunities for 'catch-up' growth, while countries that have their institutions and policies badly wrong will not only be poor but will also have low rates of growth.(Clague 1997a: 26)

This comes very close to the phenomenon of "conditional convergence".

Institutional approaches to economic growth and development can be individualized and tailored to any country. Their assumptions are too general to be contradicted by any country-specific situation, and numerous examples could be found to highlight the importance of institutions in the GCC states. While Middle Eastern Studies are marked by "grounded theory" studies until today, two important volumes on Saudi Arabia are based on institutionalist theories: in 1997, Kiren Aziz Chaudhry tried to explain the development of the institutional setting in Saudi Arabia and the Yemen (Chaudhry 1997), while Steffen Hertog recently studied Saudi Arabian institutions. (Hertog 2011)

Examples for bounded rationality can also easily be found in the GCC context. The principle of the *homo oeconomicus* is put into question by the importance of religion. (Naughton/Naughton 2000) How else could it be explained that Islam is still valued so highly, even if it may prove economically cumbersome? Why should people invent new banking products in order to avoid *ribā* (interest, usury)? From a Western perspective, Arab countries have the reputation of being marked by the paramount importance of family ties and patronage, but also unreliability and corruption, pointing to a rather weak official institutional environment. Dietz, Neumayer and de Soysa (2007) found that corruption might even be the reason why resource abundant countries invest and save much less than their wealth would allow them to save. Hence, the GCC states struggle to improve their institutional setting. Their Corruption Perception Index ratings (Transparency International 2012), for example, are better than in North Africa, Lebanon, Iraq, Syria or the Yemen. Recently, all GCC governments have aimed at attracting more foreign investment by improving their business laws and by simplifying the bureaucratic process of registering new businesses. Thus, it can be expected that NIE will be a powerful theoretical perspective for the explanation of their growth performance.

It has become state of the art of empirical studies to include institutional variables into cross-country growth regressions. Common variables are democracy, rule of law, corruption and political stability. All these variables concern the strength or weakness of institutions, and reflect the reduction of uncertainty reached in a country's institutions. (Obersteiner/Wilk 1999: 23)

It is one of the core questions of political economy as to how democratic institutions influence growth. The connectivity of economic development or growth and democracy has been analyzed in two directions: does democracy enhance economic growth and development? Or, is it the other way around, and does the achievement of a certain stage of development facilitate the sustainability of democracy? (Pourgerami 1988: 123) The latter relationship has famously been analyzed by Seymour Martin Lipset, who concluded that "the more well-to-do a nation, the greater the chances that it will sustain democracy." (Lipset 1959) Probably, this is a complex circular relationship and it would be very interesting to analyze Lipset's hypothesis in the Arab Gulf countries.

Nevertheless, this study is about how democracy influences growth only. Research on this question has produced contradictory results: those who expect democracy to have negative effects on economic growth argue that voters would push governments for redistributive measures. The latter would be granted, especially shortly before elections, because the ruler or ruling party needs electoral support. Interest groups can seek advantages, which harm the overall economic performance. (Baba 1997) Even highly inefficient measures of redistribution may be implemented at election times. Furthermore, an accumulation of such measures before an election may lead to disruptive policies. Frequent changes of government cause even more disruptive policies, and a powerful opposition may water down decisions to an extent that makes them more harmful than helpful. (Olson 1982; Nordhaus 1975; Keech 1995)

Implicitly, these arguments are based on the assumption that the state needs a certain degree of autonomy from interest groups and also voters so as to act with economic efficiency. This is probably even more important when the capitalist economic system is state-sponsored at its origins. The relationship of the state with the administration and the private sector needs to be fruitful and not only marked by self-interests. (Evans 1995: 71) However, this autonomy is not necessarily facilitated by authoritarianism or the abundance of rents accruing to the state.

In the Gulf States, inefficient redistributive policies persist despite high degrees of authoritarianism. The mere fear of their legitimacy being put into question or any potential turmoil can cause governments to throw caution to the wind. Even during times of low oil prices and budget deficits, most GCC governments did not dare to cut back welfare spending significantly. Of course, this is not valid for all authoritarian countries. A government with a different cultural background and fewer financial means has to rely more on its intelligence, intimidation and security forces.

The big difference, however, is that authoritarian redistributive states have the power to implement a long-run policy as long as they are stable and show a serious interest in the economic performance of their country. They do not need to plan from election to election, but can envisage larger long-term projects. Nevertheless, they may also face opposition which is not necessarily a parliamentary opposition.

On the other hand, institutionalists claim that democracies are more likely to have highly developed institutions. (North 1989; Knack/Keefer 1995; Kormendi/Mcguire 1985; Clague/Keefer 1997a; Wittman 1989) These institutions in turn provide information and make policy decisions more transparent and accountable. Property rights are usually more guaranteed in a democracy, and investment rates have been shown to be higher in democracies. This may also be due to a higher degree of identification between a responsible citizen and the country than a "subject" might feel for his. Moreover, the election processes is deemed more likely to produce a more competent leader.

From the Gulf perspective, I have difficulty when considering democracy as the sole "producer" of a sound institutional environment. While the institutions of the Gulf States still suffer from several problems, they are astonishingly sound compared to the rest of the region. They do not at all correlate with their degrees of authoritarianism. Moreover, dynasties can also produce competent leaders. GCC successions, either regular or by *coup d'états*, have shown a remarkable tendency to bring competent leaders to power while this is rather seldom the case for Middle Eastern elections.

Since the GCC states do not support the theoretical arguments advanced, I propose the following mechanism: institutional quality in the GCC has been improved mostly from above. Even if there have been precursors to the "councils" of today, the current crop of elections only started in the (late) 1990s – if at all. As the institutional setting had already been established there is no reason to assume that these new councils have improved this setting in an important way.Moreover, the guarantee of property rights preceded meaningful *shura* councils. There is thus little reason to assume that the recent participation has had a reasonably positive influence on economic growth. Rather, I would argue that in societies where rent seeking is widespread and where people are used to vast redistributive measures, parliaments use their power to keep their privileges. Rich merchants who have benefitted from government privileges are likely to be in parliament and do not want to see their privileges curbed. Governments who have become engaged in inefficient and costly redistribution would be deprived of their "authoritarian plus", revoking certain measures unilaterally if necessary. What is more, if their parliaments have meaningful rights, governments will be prone to "buy off" candidates with even more privileges.

This potential relationship reflects another phenomenon of the GCC states. In many respects, the governments are more progressive than large parts of their societies. Election turnouts have shown, for example, that Islamists and tribal elders have been very successful in elections. The government is much more likely to *appoint* technocrats than voters would elect them. The same is true for women. While women have been enfranchised in several Gulf States, hardly any women have been elected. It seems that even women did not vote for other women. Again, governments made sure to appoint women to some ministerial posts or to the non-elected chamber of parliament.

In a nutshell, democracy is not likely to enhance growth in traditionalist rentier societies in the short run. In the long run, however, true participatory measures may lead to more identification with a society's political process, and maybe foster a feeling of responsibility towards the economic wellbeing of the entire country, among both rulers and their constituencies.

3.3 Resource Abundance, Islam, and Arabic Culture

The preceding sections have shown that several specificities of the GCC states put common economic theories into question. In general, phenomena related to resource abundance have remained somewhat enigmatic to growth theorists who were astonished that resource booms and the resulting massive capital accumulations did not have the expected positive effects. The following chapter will investigate this issue in more detail. It will furthermore deal with the phenomenon of *'aṣabīya*[14] and with the influences of Islam on business and law.

The notion of the resource curse is used to describe the paradox that countries rich in natural resources seem to grow slower than resource-poor countries.

Although not yet fully explained, many papers claim to have found empirical evidence for a resource curse. (Gelb 1988; Auty 1990; Gylfason/Hebertsson/Zoega 1999; Sachs/Rodriguez 1999) Jeffrey Sachs and Andrew Warner controlled for a large number of additional variables and concluded that the resource curse seems to be a very robust finding. They also tested several explanations and found evidence for Dutch disease. (Sachs/Warner 2001: 833)

Today, Dutch disease seems to be exclusively a phenomenon of developing countries, but its name goes back to the Dutch experience after large natural gas discoveries in the 1960s. In the following explanations, I will rely on the seminal paper by Cordon and Neary on the Dutch disease effect. (Cordon/Neary 1982)

Cordon and Neary investigated the influence of a boom in the resource sector on other sectors of the economy in a three sector-model: one sector produced non-traded goods (e.g., services), another one manufactured traded goods, while the third sector was the natural resource sector trading in primary products. They assumed an undistorted market factor of full employment and perfectly flexible wages. (Cordon/Neary 1982: 826f)

The short explanation of Dutch disease claims that an increase of exports of natural resources leads to more foreign currency. If the surplus in foreign currency is exchanged, it raises the demand for local currency. The result is the real appreciation of the local currency. While services can become more expensive and adapt to the real appreciation, the manufacturing sector, however, will have to stick to the prices of the world market. The products of the manufacturing sector are at least partly destined for export and must not become too expensive. This makes manufacturing less profitable, and production factors (labor) will flow from the manufacturing to the resource and service sectors. Eventually this may cause a de-industrialization of the economy.

In the model of Cordon and Neary, the effect of a boom in the resource sector has two effects: a resource movement effect, and a spending effect.

The resource movement effect leads to a movement of labor into the resource sector, where high profits can be made. Both the manufacturing sector and the service sector suffer from the withdrawal of manpower. In the service sector, however, the effect is different. While the manufacturing sector is bound to the prices of the world market, the service sector can raise its prices in order to decrease the excess demand. (Cordon/Neary 1982: 830)

The spending effect assumes that the purchasing power rises and that people are likely to demand more services. Again, there is an excess demand for services, even at higher prices. However, the higher prices will lead to a higher demand for labor in the service sector. The resulting higher wages will attract more workers and lead to an increasing output of services. Because of the higher demand for services, the manufacturing sector will lose even more workers, and "Clearly, since manufacturing employment unambiguously falls, the same must also be true of that sector's output." (Cordon/Neary 1982: 831)

In the end, it is the manufacturing sector, widely considered the engine of growth, which suffers both directly and indirectly from the resource boom. Like other elegant economic models, Dutch disease relies on several assumptions that are not necessarily given in the GCC states or other resource exporters. What is more, government policies can interrupt the mechanisms of Dutch disease. Richards and Waterbury, for example, found that the Algerian manufacturing sector grew stronger during high rather than during low oil prices and gives the following explanation:

> The reason for the discrepancy between this evidence and the predictions of the [Dutch disease] theory was that Middle Eastern and other oil-exporting governments often intervened directly to counteract these relative-price shifts, usually by subsidizing their tradable goods sector. (Richards/Waterbury 2008: 15)

Besides de-industrialization, several other effects have been associated with resource booms. Rent-seeking and corruption seem to be a kind of "moral consequence" of resource windfalls. Pedro Vincente found weak empirical evidence for his supposition that corruption and natural resource windfalls are positively linked. There seem to be two different scenarios. If a country has very few problems with corruption, windfalls seem to have no influence. On the other hand, countries already suffering from corruption tend to experience an increase in corrupt practices. (Vicente 2006: 123f) This would be a useful inference if the GCC states were not outliers in his regression.[15]

Nevertheless, corruption and rent-seeking are problems in almost every society and they can be costly for economic growth. (Murphy/Shleifer/Vishny 1993; Mauro 1995; Leite/Weidmann 1999)

Illustrating a worst case scenario, Gylfason (2007: 7-11) identifies seven reasons why natural resource dependent countries underperform economically. Apart from Dutch disease he blames rent-seeking due to a massive inflow of money. Rent-seeking diverts people from fruitful economic activities, and therefore harms growth. Based on his earlier research, Gylfason adds that resource abundance lowers the incentives to invest in education and human capital. He discerns an overall trend that resource-abundant countries spend less on education than a control sample. "Awash in easy cash, they may find that education does not pay. Nations without natural resources have a smaller margin for error, and are less likely to make this mistake." (Gylfason 2001: 858)

Massive exports of natural resources might lead to currency appreciation that makes saving and investment less attractive. Growth without investment is hardly imaginable.

The fact that the revenue from natural resource exports is often invested abroad makes it less urgent to develop a well-functioning financial system. This and other factors mentioned above (the Dutch disease), in turn, make resource-dependent countries less attractive for foreign capital. Finally, Gylfason points to the problem of nonrenewable natural resources as a factor of production. If this factor becomes more and more scarce, it will limit the production output and labor will have diminishing returns.

Gylfason's economic outlook is rather bleak. Later in this study, it will become clear that many of these potential explanations are not valid for the GCC states. This calls for an explanation of why the GCC states implement policy measures when they have few incentives to do so, at least according to widespread assumptions of the resource curse.

Another question has to be considered: in some European countries and in the US, resource booms did not have negative effects on growth. They were instead seen as the real cause of their economic strength. The hypothesis to be drawn from this could be as follows: if resource booms occur in industrialized countries where institutions are sound, the rents can be used to expand the economy and they are not likely to be exported on large scale.[16] When oil was discovered in the Middle East, however, local economies and modern institutions were not yet developed. To use the arguments of the *dependistas*, colonial powers did not want them to be anything else but primary product exporters. Such a hypothesis implicitly assumes something that is not easily reflected in cross-country regressions or economic modeling: path dependence.

While many of the above mentioned theories stem from experiences in Latin America, the theory of the rentier state had the Middle East as its initial role model. The notion "rentier state" was originally used by Hossein Mahdavy to describe Pahlavi Iran in 1970. (Mahdavy 1970; Schmid 1997: 28)

Beblawi and Luciani, editors of the seminal 1987 publication *The Rentier State*, consider their concept as a part of the political economy tradition. The state and the economy are seen as closely interlinked and the concept deals with both the rentier state and the rentier economy. They define the rentier state as:

> any state that derives a substantial part of its revenue from foreign sources and under the form of rent (i.e. because specific conditions allow it to be the direct beneficiary of income derived from selling goods or services at prices well above their production costs). (Beblawi/Luciani 1987, 11)

The rentier economy is accordingly "substantially supported by expenditure from the state, while the state itself is supported from rent accruing from abroad." (Beblawi/Luciani 1987, 11)

To begin with, the notion of "rent", as used in the literature on rentier state theory, shall be explained. While economists might think of the Ricardian or Hotelling rent, rentier theory uses a broader concept. Rents are revenues that do not stem from any effort of the society. Largely, rents have not been generated by work or other production factors. This definition coincides roughly with the economic definition of rent, which is the difference between the market price of a good and its opportunity cost.[17] According to Marshall, rent is "income derived from the free gifts of nature". (Marshall 1997: 74)[18] They usually accrue to governments that can dispose of them freely. (Schmid 1991; id. 1997) The most important rents are oil and gas reserves, but also diamonds or any other natural resources can generate rents. Sometimes, even the fees a country receives for a pipeline crossing its territory are considered rents. Additionally, there are agrarian rents from tea or coffee, income from capital (interest, dividends) or payments of compensation (especially in the case of Israel). Even development aid can be considered rent. (Younger 1992)

Most countries receive rents in one of the above mentioned forms. What matters for rentier state theory, however, is the importance of the rent as government income. In most cases, one would look at the percentage of rents in the government revenue. A widely accepted threshold between "normal" states and rentier states is 40 percent. Others use the distinction of rentiers and semi-rentiers (Pawelka 1993: 110ff and 126ff) to divide the Middle East into the oil-exporting countries and the labor exporting and less oil-rich countries. One example of a semi-rentier is Egypt, which receives rents from the Suez Canal and a lot of aid from the Arab Gulf States. The Middle East has indeed developed a complex interdependent rentier-system. (Gilbar 1997: 62-73; Raffer/Salih 1992)

Rentier wealth has multiple consequences for the state and the economy. Beblawi identifies several features of states where the rentier situation prevails. Firstly, rentier states rely on a substantial *external* rent. If the rent were only internal, it could not be generated without a strong productive domestic sector. In rentier economies, however, such a sector is usually underdeveloped. (Beblawi 1987: 87)

A rentier economy is an economy where the creation of wealth centers on a small fraction of society; the rest of society is only involved in the distribution and utilization of this wealth. (Beblawi/Luciani 1987: 12) A rentier state depends on rents rather than on taxes as a source of income. It acts as distributor of rents. What is more, the government usually manages to sustain its power because it uses rents to "buy" the support of the people.

The system of rent distribution becomes problematic once the availability of rents has caused a "rentier mentality". Beblawi describes rentier mentality as follows:

> The basic assumption about the rentier mentality and that which distinguishes it from conventional economic behaviour is that it embodies a break in the work-reward causation. Reward – income or wealth – is not related to work and risk bearing, rather to chance or situation. For a rentier, reward becomes a windfall gain, an *isolated* fact, situational or accidental as against the conventional outlook where reward is integrated in a *process* as the end result of a long, systematic and

> organised production circuit. The contradiction between production and rentier ethics is, thus, glaring. (Beblawi/Luciani 1987: 52, italics as in original)

Rentier mentality accordingly leads to an attitude of waiting, rent-seeking, and networking for income. The state "allocates" the rent that accrues from abroad. Luciani distinguishes between allocation and production states. (Luciani 1987: 63-82)[19] Like Dutch disease, the rentier mentality negatively affects the manufacturing sector or, if there is no manufacturing sector deserving the name, may block its development. Politically, autocratic structures can prevail through rent distribution. This is how Martin Beck sums up the rentier concept: massive income from rents can cause a stabilization of economic inefficiency and autocratic regimes. (Beck 2007: 45)

Rentier theory has its critics too. Several resource rich countries seem to contradict rentier theory by either developing a viable, sustainable economy or by undergoing democratic reform. Mostly, intervening variables can be detected in those cases, pointing to the possibility of path dependence: The rentier state is obviously not a necessary consequence of resource abundance. The existence of counter-factual cases, however, does not make rentier theory obsolete. It is still one of the main theoretical approaches towards resource abundant states, although there recently have been modifications and enlargements. In a detailed case study of Saudi Arabia's institutions, for instance, Steffen Hertog presents the concept of "segmented clientelism" (Hertog 2005 and 2011) to describe how rentier structures are deeply rooted in Saudi Arabian society, economy, and bureaucracy but to different degrees. Rather than a contradiction, this is a long needed elaboration of rentier state theory on a deeper, more detailed level.

In another recent study, Sean Foley (2010) argues that the Arab Gulf states have been too long viewed from the rentier angle only. While he certainly has a point in saying that the GCC states were influenced like any other country by international events, his analysis has unintentionally confirmed the difference oil wealth makes in dealing with these influences.[20]

The GCC states are widely considered the examples par excellence of the rentier state. Not long ago, their revenue consisted up to 95 percent of rents. Rents are not only the power basis of the ruling families, but furthermore shaped the borders of their realms. When it comes to rentier mentality, it is widely assumed to be widespread in the GCC states, and employment statistics have to be interpreted with care. A considerable percentage of the nationals employed seldom show up to work, or "works" only a few hours between long lunch breaks and long prayer times. However, low work motivation may also be due to other factors, such as strongly hierarchical business structures that often reduce the individual to a mere recipient of orders, even if he or she might have more expertise than the director who often draws his authority from age and position only. Slow and inefficient processes, a high power distance, and the lack of open and constructive feedback in societies where critique is avoided in order to keep conversations harmonious often preclude employees from getting a sense of ownership of their work. (Hofstede 1980; Hamaiza 2015; Hsu 2015) Decades of rentierism have nourished a set of negative work behaviors and have given traditional hierarchies the chance to survive. These structures have survived rentier mentality and young highly motivated individuals may be unable to change them.

At the beginning of chapter two, I already mentioned Islam and tribal affiliations as an important feature of the GCC states today. The importance of these features is often underrated as their effect on the economy is often revealed by small details only. It will soon become clear that Islam and non-religious social factors cannot be separated. Despite widespread Islamic appearances, many tribal aspects have not been eradicated.

It is also important to note that Islam as an umbrella term for different movements can hardly be put into a *direct* relationship with economic growth. Islam can be, and is interpreted and lived, in several ways, and each way might have different implications for the economy. This said, Islam must not be reduced to a theoretical ideal. An example of this danger can be found in the statement by Maxime Rodinson, one of the early writers on Islam and economics:

> la religion musulmane n'avait jamais élevé d'objection contre le mode de production capitaliste et que ses prescriptions n'avaient pas constitué une entrave au développement de ce mode de production dans la direction de la constitution d'une formation économico-sociale capitaliste. (Rodinson 1966: 148f)

From his socialist background, Rodinson made colonialism and imperialism responsible for the economic decline in the Islamic world. If Rodinson were completely right on this issue, I could refer to dependency theory and this chapter would be superfluous. However, the colonialist and imperialist penetration of the Middle East was only possible because it already lagged behind. As an important feature of Middle Eastern societies, one has to at least consider Islam as a potential reason for this. (Kuran 1997: 46ff) Nevertheless, Rodinson's opinion implicitly points to an often neglected fact: almost the entire world lags behind the West economically.

Islam itself has not precluded the development of capitalism but has been one of the factors that have affected social behavior, norms, and institutions and, ultimately, the economy. Moreover, both the Islamic reality and the non-Islamic environment have undergone changes. It is obvious that the conclusion of this chapter cannot be a clear-cut "Islam impedes" or "Islam enhances growth". Statements of this kind are often made and resulted in both biased condemnations as well as biased appraisals. Both Muslim and Western writers got caught in the trap of taking a few aspects and a certain period of the history of Islam in order to identify it as being of economic advantage or disadvantage.

Among the "disadvantagists", Max Weber's remarks on Islam are a convenient but much critized starting point. In *The Protestant Ethic and the Spirit of Capitalism*, Weber singles out the Protestant belief in predestination as a driving force of the rise of capitalism in Western Europe. The Protestant Christian considered worldly success as a sign to be among the elected people of God. Because Protestants are directly responsible for their life before God, they should try to be honest and self-sufficient. Inner-worldly asceticism would make them reinvest their profits instead of spending them on luxururies. Equally important is the rationalization of religion and the world. Sacraments or priests have no magical influences any more and religion is now a matter of cause and effect: a good life leads to election after death. (Turner 1998; Sukidi 2006; Huff/Schluchter 1999)

Weber considered Islam as the opposite of Protestantism. Islam would be marked by predetermination in the sense of *life* being predetermined. There is no double-edge possibility of either election or damnation after death, and therefore, there is no struggle for election.

In Weber's view, Islam lacks the idea of life being a practical test for a life to come. The result would be fatalism. (Weber 2006: 13f) The anecdotal *IBM*-strategy – *Inshallah, Boukra, Maleesh* (If God wills, tomorrow or later, it doesn't matter) – widespread in many guides to doing business in the Arab world, demonstrates how many Westerners still perceive the "typical" Middle Eastener as a "fatalist". However, this fails to explain how fatalistic elements in Islam become so important while they did not in other religions. After all, most religions contain elements of fatalism.

Famously criticized by Edward Said, many "Orientalists" held the view that, since Islam would never change, modernization would only be possible without Islam. (Kuran 1997: 50) Muslim authors like Niyazi Berkes (1964) also tended towards such explanations by stating that Atatürk's reforms could be implemented only by a retreat from Islam. (Kuran 1997: 52) As recently as in 1973, I. Adelman and C. Morris ranked religions according to their effects on economic behavior. Islamic countries received a "B" standing for "countries in which the predominant religion promotes moderately fatalistic attitudes towards man's capacity to alter his destiny." (Adelman/Morris 1973: 39) In 2003, Guiso, Sapienza, and Zingales interpreted data from the world value survey and compared economically "useful" attitudes across religions. They found that these attitudes are valued highly in predominantly Protestant countries while they are of comparatively little importance in predominantly Muslim countries. Muslims ranked behind Jews, Catholics and Hindus. (Guiso/Sapienza/Zingales 2003)

This "thesis of economic disadvantage" contrasts sharply with the literature on Islamic economics. The origins of this literature can be traced back to the 1950s and 1960s when the world was divided into East and West or communist and capitalist countries. Several Muslim writers portrayed Islam as a third way. Muslims could not accept communism – a concept that denied the existence of God – nor could they approve of unrestricted capitalism. The first literature thus focused on what Islamic economies should *not* be. Several authors propagated a kind of Islamic socialism, arguing that Islam had already propagated socialist principles before socialism was invented. (Philipp 1990: 120) Others emphasized the Islamic guarantee of private property and advocated an alleviated capitalism. Most claim that a just Islamic economic order had already been established in early Islam and that the economic decline was due to the neglect of Islamic economic principles. Later, colonialism and imperialism added to this neglect. The Pakistani author Raguibuz Zaman, for example, states that

> [T]he earlier centuries of Islam were witnesses to the efficient functioning of the Islamic economic system and an analysis of the workability of this system in modern times will draw upon this experience. (Zaman 1980: 80)

As a general conclusion from several Muslim writers on Islamic economics, Thomas Philipp concludes that

> [N]ot one of the authors asks a critical or analytical question from history. The past is what it is asserted to be. ... the Muslim economist treats an ideal order projected into the past as a historical fact and then draws from there legitimization for his own theoretical endeavors. (Philipp 1990: 126)

Before I turn to a hopefully more balanced approach towards Islam and economic performance, let me point to an important link to the Gulf States. Their economic development was booming exactly at the time when the field of Islamic economics was also rapidly growing. In the 1960s and 1970s, the ban on interest and the introduction of an alms tax (*zakat*) were discussed. (Nienhaus 2006) More than other Arab or Islamic states therefore, the GCC states had the opportunity to consider the economic aspects of Islam.

To my knowledge, there is no theory linking aspects of Islamic economics such as interest-free banking or *zakat* to good or bad economic performance. Literature on these issues is still confined to discussing diverging attitudes towards Islamic banking or *zakat* and how they have been implemented.

There is, however, a very elaborate discussion meanwhile on why the Islamic world fell behind the West economically. In this discussion, several institutions have been identified as cumbersome. While efficient during the Middle Ages, these institutions became a burden in a changing, increasingly impersonal economy. In the following chapter these factors will be discussed in order to critically examine if they are still relevant today.

While it was his "fatalism" argument that gained most attention, Weber also adduced economic and political arguments. Bryan S. Turner summarizes Weber's view on Islam and capitalism as follows:

> it was the patrimonial nature of Muslim political institutions which precluded the emergence of capitalist preconditions, namely rational law, a free labour market, autonomous cities, a money economy, and a bourgeois class. (Turner 1998: 2)

Also Timur Kuran acknowledges the merit of Weber's identification of some of the sources of the economic decline of the Islamic world. "Max Weber invoked several factors as explanations: the inflexibility of Islamic law, the failure to achieve legal formalization, arbitrary personal rule, and a lack of ascetic streak fueling 'salvation anxiety'." (Kuran 2008: 41f) All these arguments are valid to a certain extent but it is important to keep in mind that they do not answer the main reason. Why have all these institutions – an "inflexible" Islamic law and arbitrary political rule – turned out to be so persistent?

The initial Islamic expansion was very successful. Muslim merchants were in control of the most important trading routes at that time. The conquest and Islamization of hitherto "pagan" territories was a religious task and combined very well with trade. Islamic law contributed to the success of Muslim traders by providing for what was then a modern contract law. (Hassan 2011)

Between the 7th until the 10th century, Islamic law was elaborated although not uniformly codified. Religious scholars devised concepts of the Koran and the *sunna*. This was necessary because rather than clear rules for everyday life, Koran and *sunna* provided moral principles. Apart from the primary sources of law (Koran and *sunna*), there are several "instruments" of jurisprudence that were used to draw conclusions. Apart from the highly esteemed *ijmā'* or consensus of all scholars (Nienhaus 1982: 82), secondary sources of law are rules resulting from *qiyās* (analogy). Through *qiyās*, cases without a clear precedent in Koran or s*unna* were treated analogously to other established cases, if they were sufficiently similar. (Schacht 1965: 30 and 38)

Of course, there were issues neither mentioned in the Koran and *sunna* nor settled by *ijmā'*. Here, Muslim scholars (so called *mujtahids*) had to decide according to their personal *opinion* (*ra'y* in Arabic). This is called *ijtihād* which Wael al-Hallaq describes as follows:

> *Ra'y al-ijtihād* [...] meant the intellectual activity or the reasoning of the legal scholar whose sources of knowledge are materials endowed with religious (or quasi-religious) authority. At a time when the textual sources of religion were not yet established and when their controlling authority was far from exclusive, such practices as *ra'y* would not have been censured. In fact, by the eighth decade of the first [Islamic] century, the term was used to indicate sound and considered opinion. (Hallaq 1997: 15)

According to Joseph Schacht, the initial acceptance of *ra'y al-ijtihād* declined and rulings had to be based on *taqlīd* (imitation) of what scholars had already written. This rupture is often refered to as the "closing of the gate of *ijtihād*" around 900 AD. (Schacht 1965: 69-75) Other scholars argue that it continued in later centuries and some, like Hallaq, even claim that the door of *ijtihād* has never been closed, "neither in theory nor in practice." (Hallaq 1984: 4; 37) There probably has not been such a net rupture with *ijtihād*, and Schacht admits that independent reasoning continued to be applied. (Schacht 1965: 71) All four Sunni orthodox schools of law showed

> manifestations of original thought ... But this original thought could express itself freely in nothing more than abstract systematic constructions which affected neither the established decisions of positive law nor the classical doctrine of *uṣūl al-fiqh.* (Schacht 1965: 72)

The discussion about *ijtihād* is very complex and the fact that the word *ijtihād* has changed its meaning over time only adds further complexity. Even if the formal development of Islamic Law stagnated in the 10th century, there has always been the need for decisions on matters without precedent. In those cases, Muslim judges often used casuistry or *ḥiyal.* Casuistry helped to circumvent cumbersome bans and rules of Islamic law but they did not change the existing positive law. The main focus of Islamic law remained the primary sources and there was no "system of granting authority to precedents". (Kuran 2008: 28) By casuistry, a current problem could be overcome, but casuistry could not change institutions.

These legal devices were economically relevant. They were used to allow for the charging of interest by merchants or moneylenders through the notion of a double sale. (Schacht 1965: 78f) However, this gimmick did not provide a lasting solution to the problem of the economic need to give money and time a price by charging interest. Today, *murābaḥa* (a resale instrument in Islamic finance) is still used widely but not acknowledged by many religious scholars, and basically all of them claim that instruments of real risk and profit sharing are preferable. (Kuran 2004: 8; 10; 108)[21]

These differences between theory and practical reality can be identified as a "behavioral lag" (John Waterbury). Islamic law was neither relevant for "modern" cases, nor was it allowed to extend them, so people went on to find their own solutions. The gap between reality and the ideal widened. (Waterbury 1970)

As a side note, a similar "lag" can be detected with regard to the Arabic language. As the language of the Holy Koran, *'Arabīya* had attained a status of something accomplished and unchangeable. During the Middle Ages, the Arab language functioned well as *lingua franca,* but increasingly the fact that Arabic was tied to the Koran and used by the Ottomans for official documents led to a double situation. Koranic Arabic was left (almost) untouched, but since there was a need for new expressions, dialects developed. These dialects – like the *ḥiyal* – were never institutionalized or formally written down. There is even a philological committee to defend the pure Arabic against newer versions that, to be sure, have developed in the science and business environments. (Tibi 1985: 99-130)

During the Middle Ages, however, Arabic and Islamic law were still a huge advantage. The inheritance system was elaborated and commercial contract law were well established, together with *qādīs* (judges) to hear cases. This proved to be highly effective. In territories without written law, it became attractive to convert to Islam in order to become a member of Islamic business circles, where certain principles were guaranteed. Commercial law was sufficient to organize short-term business partnerships between two partners. In the personalized trade of the Middle Ages, the Islamic system was thus not only sufficient, but marked by a hitherto unknown coherence. From West Africa to India, commercial law differed only slightly according to the school of law which prevailed. (Kuran 2008: 6f)

By the end of the 17th century at the latest, Europe prepared for capitalism: the banking system became more sophisticated, corporations were founded and the rule of law improved. This, in turn, paved the way for expensive investments into new technologies. The size and durability of new companies allowed for a division of labor and economies of scale. (Kuran 2008: 10)

In the Middle East, Islamic law did not provide for legal personality or corporations and without influence from Europe this rule was not to be broken (Kuran 2008; Kuran 2005) There were daring and innovative entrepreneurs. The 17th century, for example, saw coffee traders (Kuran 2008: 11) although coffee was then considered intoxicating and condemned by many Muslim clerics. These entrepreneurs were pioneers but they were not able to change the system. Their work relied on short business partnerships, usually with people whom they personally knew. When one of the partners died, his or her fortune was often divided into many small parts to serve all heirs. This usually meant the end of the partnership and it made sense to keep projects short. As a result, businesses remained relatively small. (Kuran 2008: 15)

Both weak contract law and Islamic inheritance law precluded the advent of big companies that would have been able to compete with European ones to benefit from economies of scale.

However, Muslims were innovative in circumventing Islamic inheritance law by creating *awqāf* (plural of *waqf*, religious foundations). The *waqf* is neither directly mentioned in the Koran nor restricted to Muslims but is in the spirit of the duty of almsgiving. People having legally acquired a certain possession dedicate it as *waqf* to the well-being of society. This property can become a school, a public library, a soup kitchen, a hospital or a public well, but also the *usufruct* from the territory can be used, for example, to pay the salaries of doctors or teachers.

The popularity of *awqāf* is usually attributed to the fact that property could be protected from both seizure by the often-expropriating state, and from parceling property in case of many heirs. Islamic states had an interest in *awqāf* because these religious foundations provided public goods which the state did not provide. On the other hand, it was very attractive for a private individual to turn his or her property into a *waqf* and to make his or her favorite relative *de facto* heir.

The consequences are less clear. While *awqāf* provided a lot of public goods and contributed to economic well-being, the meaning of the word *waqf* ("stopped, prevented", from being sold, inherited, or rented out) already contains an element of stagnation. Once a territory or building was designated for a certain purpose it was very difficult to change this even if its purpose had become obsolete or was not lucrative any more. The economic conclusion could be that *awqāf* bound capital that could have been used more efficiently elsewhere. (Kuran 2008: 38f; Kuran 2001)

Another issue was cash *waqf*. Even if highly disputed, moneylenders turned their fortunes into cash *awqāf* and went on to then lend it to their clients. A moneylender usually lent money to the people he knew, but his or her fortune was not sufficient for huge commercial loans, necessary for investments into technology. This problem could have been overcome by the merger of several moneylenders, but then, the basis for business transactions in the Middle East – mutual trust – would not have been sufficient anymore. Viable institutions and laws would have been necessary to provide security to all parties involved in a big finance deal, but these institutions were not available. (Kuran 2008: 29)

Today, *awqāf* are still a very popular way of giving charity in Arab countries, and all GCC states have a rich heritage of *awqāf*. Usually, there is even a ministry for religious endowments, sometimes combined with the the ministry of Islamic affairs. In some countries, attempts have been made to register all the *awqāf*. (Benthall 2002)

The conclusion from these few examples of the medieval Islamic economy is that there were always innovations, such as *waqf*, new products or casuistry to circumvent certain statements of Islamic law. The argument that innovations (*bid'a*) were always to be rejected in Islam does not hold as strongly as it is often affirmed. In some cases, for example when explaining why the printing press was used only so late in the Ottoman Empire, the real obstacle was not the clerics but other factors, such as too little demand for printed books and the opposition of the guild of writers. (Kuran 2008: 21-24)

Going back to Weber's argument, another problem was the dual legal system that evolved in many Middle Eastern states. While the caliph was entitled to apply secular law, he was not given legislative functions in the matters treated by Koran and *sunna*. Although *sharia* law "had to resign an ever-increasing sphere to practice and custom" (Schacht 1965: 77), thereby enlarging the scope of manoeuvre of the ruler, the latter could always be challenged by rulings of clerics and often shied away from important economic changes. (Kuran 2008: 32). In general, the government often opted for short-lived measures if it needed more taxes, but avoided and prevented far-reaching innovations, even if the latter stimulated growth and therefore pools for tax-raising. (Kuran 2008: 35)

There are essentially two explanations for this lack of change. First, there is the famous argument by Bernard Lewis, who observed that interest by Middle Easterners in Europe was practically absent until the 20th century. There were few translations of European books, no grammar books of any European language, and few Ottomans traveled to Europe. During the various wars, mostly in Eastern Europe, one observed that Europe had developed better weaponry but one failed to identify the underlying causes: a better schooling system that was not based on memorizsation, and a culture of "rational science". In the Ottoman Empire, by contrast, a "mistrust of novelty" would reign. Instead of analyzing "what went wrong", to use Bernhard Lewis' words, many Muslim scholars would look for a supposed golden age in the past.

Islam was among the reasons for this "mistrust of novelty" since the *ulema* (*'ulamā'*) discouraged innovation by reiterating that Islam was already perfect and that novelties could only lead to its decline. Bernhard Lewis may also be right in saying that Islam was deemed superior to the *dār al-ḥarb* and that it was difficult to accept that something good could come from the land of the infidels. Rather, if the *dār al-islām* had economically been overtaken by the "land of the infidels", this was because the former had been subject to (bad) innovations which distanced it from the Islamic ideal. This way of reasoning also facilitated an educational system which did not encourage autonomous thinking but rather the repetition of dead knowledge. (Lewis 1982; Lewis 1993) The medieval philosophers – al-Farabi (around 870-950), Ibn Sina (980-1037) or Ibn Rushd (1126-1198) – had failed to establish a philosophy (*falsafa*) that was independent from theology (*kalām*). Instead, the notion of science took the meaning of religious scholarship and "rational science" almost came to an end. Already in the 14th century, Ibn Khaldūn (1332-1406 AD) had observed the decline of sciences in Islamic regions, while he saw them thriving in the "land of Rome":

> Later on, civilizing activity stopped in the Maghrib and in Spain. The sciences decreased with the decrease of civilization. As a consequence, scientific activity disappeared there, save for a few remnants that may be found among scattered individuals and that are controlled by orthodox religious scholars. We hear that the intellectual sciences are still amply represented among the inhabitants of the [non-Arab] East. [...] We hear now that the philosophical sciences are greatly cultivated in the land of Rome and along the adjacent northern shore of the country of the European Christians. They are said to be studied there again and to be taught in numerous classes. (Ibn Khaldun 1967: 375)

The fact that religious scholarship had replaced sciences for a very long time is well illustrated by Rifā'a R. Ṭahṭāwī (1801-1873), one of the first Ottomans to study in Paris. He observed that science meant something entirely different in France. Researchers specialized in certain fields in order to explain facts and to bring about useful things while priests were only scholars of religion. (Tibi 1985: 111-113)

Bernard Lewis' argument is supported by much other research, among others by the Syrian Bassam Tibi who confirms this mistrust of novelty. He cites not only historical examples, but also quotes contemporary fundamentalists who clearly express their view that Islam is already perfect and that there is no need for modernization. (Tibi 1985: 96) While Lewis' arguments stand on firm ground, there are still a lot of points where critique may be used to attack the aforementioned explanation. Particularly, Bernhard Lewis, in explaining why Muslims noticed developments in Europe only late, seems to restrict himself to arguments supporting his view. To him, only knowledge about Europe counts, while he does not acknowledge that Muslims often learned from non-Europeans and knew non-European languages. These languages may have been more important for them than European languages. Furthermore, Muslims lived under non-Muslim rule, for example in China. (Bulliet 1983)

While Islam has often played a role in the aforementioned mechanisms of "stagnation", there are further explanations that are neither based on Islam nor on a specific interpretation of Islam but on prevailing pre-Islamic social features. This is the case of *communalist* or *collectivist*[22] societies:

> The essence of communalism is that one's rights and duties spring from one's status in the community; the 'good' is the common good of society, which is generally small and considered largely self-contained. (Kuran 2004: 139)

By contrast, an individualistic society

> ...provides broad personal freedoms concerning activities, occupations, beliefs, and responsibilities; and, in approving of self-determined thought and conduct, it seeks to create the conditions that maximize the enjoyment of such freedom. (Kuran 2004: 139, based on Oakeshott 1958/1993: 18-21)

A. Greif (1994) analyzed organizational patterns and interactions of Genoese and Maghribi traders in the Mediterranean in the 11th and 12th century and found different structures: the collectivist (communalist) Maghribi traders featured horizontal economic interactions, clear segregation, and a stable wealth distribution. Economic and social interaction took place in a small group, which facilitated intra-group communication. Problems were dissolved within the group and no third party was needed. Whoever deviated from the rules could be penalized, for example, by exclusion.

The Genoese traders showed the beginnings of an individualistic society. Economic and social integration was vertical and the dependence of the individual upon one group was weak. Potential sanctions against one member were therefore not as threatening as in the close-knit Maghribi groups of traders. In order to enforce rules, a third formal party – a legal institutions or court – was needed.

In history, the Genoese traders drove the Maghribi traders out of the Mediterranean. Today most of the developing countries, among them all predominantly Muslim countries, are "collectivist", whereas the developed West is "individualist". In the former, self-expression and secular-rational values are rated considerably lower than in Western countries. Richer Muslim countries show a higher propensity towards self-expression values, which, however, still remain much below the propensity in developed countries. (Inglehard 2005)

Individualistic societies have clear advantages in the current economic world order. Although one may argue as to what came first, individualistic societies need formal "second parties" that can enforce certain rules, and thus they require an institutional environment that facilitates anonymous exchange. Any individual is guaranteed rights that do not depend on group adherence. A merchant can look for an employee outside of his family or clan and might find someone with better skills. Additionally, people in individualistic societies feel less social pressure to adapt to norms, and are more prone to innovations or new methods of entrepreneurship, which might, at least at the beginning, break with traditions or threaten the existence of certain other groups or professions. (Greif 1994: 942ff) On the contrary, strong social beliefs in a communalistic society can preclude innovation. Not because of a conviction that innovation is bad but because of the respect of social norms. Although people may not agree in private, they may defend in public. Kuran calls this phenomenon *preference falsification*.

Collectivistic societies are less prone to rely on the government for services or legal actions. Many problems are dealt with within the family or clan without interference from third "official" parties.

While the Koran contains many meritocratic principles and praises those who work hard, Islam has failed to abolish tribal structures where personal status depends on blood and family ties. Muhammad wanted the *umma*, the community of believers, as well as *sharia* to take over the protective role of the tribe. (Rodinson 2002: 228) However, while Muhammad could establish Islam as a part of the Arab identity, tribal identity was not eradicated. This importance of tribal affiliation or *'aṣabīya* ("spirit of kinship") was described by Ibn Khaldun, who considered it positive:

> Compassion and affection for one's blood relations and relatives exist in human nature as something God put into the hearts of men. It makes for mutual support and aid and increases the fear felt by the enemy. (Ibn Khaldun 1967: 98)

In this context the social consequences of *'aṣabīya* are important. If blood or patron and client ties are the most important ties, then people will work to the benefit of their own group and will in turn expect that they benefit from other members of their group. Politically, this often means that people vote for the leader of their *'aṣabīya* group and not for the candidate with the best abilities. Party programs are obsolete because the person is the program.

As for economic implications, another important point would be that people might look for a job using their *'aṣabīya* ties. Competing for a job, those with the better "*wasṭa*" (connection) would get it, even if they are less skilled. A highly skilled job-seeker without *'aṣabīya* may not even get a chance to prove his or her abilities.[23] Furthermore, the leader of a *'aṣabīya* comes to power because he (and seldom she) offers his or her supporters something in return. Of course, charisma is required as is also a common ideology (blood ties, religious beliefs, ideological ties), but such a leader also has a clear interest in providing as many fellows as possible with a job. Even if a company or public administrator might do better with few, highly skilled employees, they might end up with many low skilled ones. *'Aṣabīya* ties may thus hamper efficiency.

The Role of Women in Arab Societies

There is another source of inefficiency in the GCC societies that has attracted a lot of attention recently. Women have come to outperform their male peers in education but are underrepresented by far in the labor force. Western media mostly points to restricted rights of women in the Arab world, and have singled out the *abāyas*, *ḥijābs* and *niqābs*, and driving bans and gender segregation at the workplace, as reasons. While women indeed face several cumbersome restrictions, the wealthier among them, however, lead privileged lifes with maids, cooks, and drivers doing their work, which makes it easier for them to work than for any Western woman. It is hard to tell if the decision to take a job is shaped more by personal preferences or by outside factors. The same, however, holds true for their male counterparts, and on average women are considered more eager to work.

The link between women's status in society and the economy can be described as follows: firstly, education of women is important, because even if they do not work themselves, they educate their children. Secondly, women contribute to the labor force and they have special skills which enrich society enormously. Thirdly, if women study and work, they usually have fewer children. (Barro 1997: 13 and 119; Griffin 1999: 133) This not only alleviates problems of population growth but also ensures a better education of children, as people having fewer children can invest more in each of them. And fourthly, even if women are allowed to work but only under several restrictions, this might raise costs and lower their chances of being employed and promoted. In societies where rentier mentality reigns women can be even more important for the economy. For them, studying and working comes along with new liberty. Having a job or studying justifies their leaving home and give them opportunities they would not have without.[24] In general, younger and educated women especially have good reasons to be ambitious, because they have an interest in change and in gaining more rights. To the contrary, men lose their predominance, which the prevailing interpretation of Islam as well as tribal bedouin society has given to them. Working for change may be less of a personal goal for them. As Jones (2015) put it, "in many respects working women are ready for the workplace, but the workplace itself may not be ready for them."

It seems to be obvious that Islam accounts at least partly for the low percentage of women in the labor force and for the still comparatively high birth rates in the GCC states. Nevertheless, when it comes to international rankings such as the Gender Inequality Index (a component of the HDI), interestingly the GCC, apart from Qatar, receives rather good values. Even Saudi Arabia, which the Western media seems to have singled out as the worst place for women to live, reached rank 56 in 2013, which is better than most South American countries. The low percentage of women in the labor force, therefore, does not seem to be too much a result of inequal access.

Recent research points to the role oil wealth may play in this regard. Michael Ross (2008) argues that oil is responsible for the little progress of women in the labor market. Indeed, he has many valid points: arguing with Dutch disease theory, Ross states that higher income in the non-tradables sector has the effect of higher wages for men, thus increasing the household income, and decreasing the incentives for women to work. On the other hand, higher wages should be an incentive for women to work. This incentive, however, could be lowered by massive government transfers to the population. In several countries, women particularly work in the tradable sector (cloth and clothing), which can be especially affected by Dutch disease, thus offering lower wages and fewer incentives. Ross empirically demonstrates the correlation of income from oil and lower percentages of women in the labor force. Islam, on the contrary, seems to be insignificant.

There are, however, shortcomings in his analysis: the use of a dummy variable for Islam (more than 50 percent of the population are Muslim) is especially problematic because it does not take into account that women have different rights in different Islamic countries. Furthermore, he does not ask why the percentage of women in the labor force before the oil windfalls was already low.

"Ross' argument is compelling but dilutes the powerful influence of Islam on Arab culture – which predates the discovery of oil by several *millennia* – a crucial key to understanding the status of women in contemporary Arab societies." (Al-Nasr 2009: 2) concludes Tofol al-Nasr, herself from Qatar, in a critique of Ross' article.

3.4 Summary: Theoretical Approaches

The preceding sections have revealed that general growth theory is not enough to explain economic growth or a failure to diversify in the GCC states. Even if capital, population growth, and human capital are crucial factors, the GCC states clearly deviate from the underlying assumptions of the model. The fact that capital is not the main constraint and that human capital is imported on a large scale questions the very foundations of these growth models.

Development theories were not tailored to the GCC states, but they nevertheless better reflect the reality in the GCC as developing states. The idea of having a comparative advantage in energy-intensive industries has played a huge role in all diversification strategies. Comparative advantage was the reason why oil and gas – as well as their down- and upstream sectors – were considered the leading sectors for governmental development planning. Big Push theories and unbalanced growth theories are especially relevant for development strategies in the GCC states.

There may also be some truth to the dependency models, but again, the assumed mechanism is not an accurate picture of the Gulf States. Oil gives them so much power that it is not possible to degrade GCC states to a "periphery", and in real terms oil and gas have become more expensive. Nevertheless, import substitution has been one of the (unmet) targets of diversification strategies.

The main advantage of institutionalist approaches to development is their universality. Institutionalist approaches do not assume a certain process of how development takes place or how it is constrained by a special constellation of factors. They do not aim at explaining the macro-development of a country, but look at a lower level, and "micro-mechanisms." Institutionalist approaches are concerned with rule of law, corruption, and in general the quality of business-relevant institutions, democratic institutions included. A short review of the literature has shown that it is by no means clear whether democratic institutions drive or impede growth. Rather, the ideal would be a government with "embedded autonomy" and it seems that this autonomy can also be reached in autocracies when other circumstances are favorable.

The specificities of resource abundant states have given rise to a vast literature on how resource wealth could affect the economy. While it seems likely that the resource curse (if it exists) could be explained by a lack of incentives to diversify, Dutch disease is a model that predicts several mechanisms on condition that several assumptions are fulfilled. However, massive government involvement into the economies that characterize the GCC states questions these assumptions. Currency appreciation can be also be controlled by government measures. Furthermore, there was scarcely a real manufacturing sector in the GCC states when the oil windfalls started.

Rather, *rentier* state theory locates the problems in society, the administration and the government; again, there is no sophisticated modeling but rather a sum of potential negative influences on the economy and the political system. *Rentier* mentality and rent-seeking provide theoretical explanations of human capital and migrant labor problems that are unexplained by common growth theory.

Islam is one of the most important factors shaping the Gulf societies. This influence is not necessarily negative: Islamic banking, for example, has been a booming market. As for potentially damaging factors, the Islamic influence on the institutional setting has to be assessed from case to case.

Although Islamism is widely believed to be unIslamic, Islam has played a role in its development. Some have taken on ideas of early *salafists*, who claimed that the return to the Islamic principles of their forefathers (*salaf*) during the lifetime of Muhammad, was the right solution for the Islamic world to regain its former glory. Today, *salafi* thoughts, it seems, have even increased in importance. They have influenced fundamentalists and terrorists. Thus, the conflict between modernizers and *salafists* shapes social and political life in all GCC states up to today.

Collectivism is an important feature of all Arab societies. Their collectivist nature explains much of their economic inefficiency. The importance of family ties and tribal affiliations hinders the emergence of a real civil society, based on chosen links and of meritocratic principles in the economy and even administration.

4

Methodology

This chapter will further elaborate the methodological framework of this study. Apart from the qualitative case studies, two methods will be used: Fuzzy Set Qualitative Comparative Analysis and regression analysis (OLS), the common method growth and development economists use, although time series or panel data analysis has become more widespread.

Additionally, the question of growth in resource abundant countries has been the object of many regressions. Scholars such as Jeffrey Sachs and Andrew Warner (Sachs/ Rodriguez 1999; Sachs/Warner 1995a/1997; Sachs/Warner 1997; Sachs/Warner 2001; Sachs/Humphreys/Stiglitz 2007) or Eric Neumayer (2004) have tested the resource curse hypothesis multiple times and found supporting evidence. Both Sachs and Neumayer used resource dependence measures for their analysis. Other scholars have changed the models and the operationalization of variables. The significance of resource "abundance" variables has been put into question. Brunnschweiler, for example, found a positive correlation of resource abundance measures with growth, significant even when testing for institutions. Brunnschweiler also included regional dummies, while only the dummy for the Middle East and Africa was significant, and negatively correlated to growth. Using these dummies, Brunnschweiler could introduce a kind of regional path dependence in a regression. However, the dependent variable was the normal growth rate and not the non-oil growth rate. (Brunnschweiler 2008; Kropf 2010)

After decades of regressions, resource wealth seems to be one of the ambiguous variables and it seems unlikely that this kind of analysis will provide a clear answer. Several authors have already pointed to the probability that resource wealth and its effect on the economy or on democracy are part of a path dependent process. (Brunnschweiler 2008; Robinson/Torvik/Verdier 2006) Smith (2005) for example suggests that independent elites are a precondition for democratic transitions in resource abundance states. Nevertheless, I will start my analysis with a regression to contrast this approach and the common operationalization of the variables. To guarantee the comparability of the methods, I need to adapt the hitherto-used models to my approach as I cannot solely rely on previous models.

To highlight the down- and upsides of each method, the regression will be the starting point of a multi-method appraoch, which encompasses a large n-sample, two small n-samples, and the case studies of the GCC states given in chapter two. This triangulation will also allow for a learning process from method to method.[1]

The main focus, however, is the small sample of GCC countries, and as the regression will demonstrate, their cases almost disappear in the larger sample of 181 countries. Earlier in this study, the chapter on case studies provided a detailed account of the single countries but without a systematic comparison. Six cases are too many to be compared unsystematically and too few for a regression on their own. Hence a systematic comparison calls for another method, which this chapter will deal with: Fuzzy set Qualitative Comparative Analysis (FsQCA).[2] It will also deal with dependent and independent variables (for the regression) or outcome and conditions, their operationalization, as well as data sources.

4.1 Fuzzy Set Qualitative Comparative Analysis

Qualitative Comparative Analysis (QCA) is an umbrella term for two different methods: crisp set QCA and fuzzy set QCA.[3]

QCA is a relatively young method, stemming, on the one hand, from the attempt to bridge the gap between qualitative and quantitative research methods. One the other hand, there is a lack of methods for small-n analysis. To my knowledge, only a few researchers have thus far applied this method to Middle Eastern countries. Philip N. Howard (2010) analyzed the role of information and communication technologies for democratic transitions or entrenchments in 75 countries containing important Muslim minorities. (Howard 2010; Howard/Hussain: 2013) Interestingly, a byproduct of his study was a confirmation of the negative effects of resource dependence on these transitions. (Howard 2010: 25)

There are good reasons to restrict a sample to those countries belonging to a certain organization or sharing at least some common features such as the EU or the OECD. Earlier, the analyst was usually faced with a choice between focusing on very few (one, two, three, but certainly not more than four) countries, or working with much larger samples, starting at about 70 cases. (Ragin 2000: 25) QCA fills this gap and is suitable for samples from 5 to 45 cases.

One of the main characteristics of QCA is the case-oriented rather than variable-oriented approach. In QCA, every case is important because it may lead to a new *possible path* to reach a certain outcome. Outliers, like the GCC states, are not excluded but stand for potentially important alternative paths.

The difference between case-oriented and variable-oriented approaches manifests in another feature of QCA. Regressions and other quantitative methods often concentrate on how much one variable contributes to the explanation of a certain outcome (Mahoney/Goertz 2006: 230) or, to use Ragin's words, concentrate on the net-effect (Ragin 2006) of one variable. The result of a regression may be that variable A contributes 15 percent to a certain rise of the dependent variable, while variable B contributes only 5 percent. QCA, however, assumes that the effect of one variable or, as Ragin would prefer, of one "condition"[4] on the outcome must never be regarded as isolated. There is not only a correlation between conditions and the outcome, but also among the conditions. This is why QCA focuses on *configurations* of conditions.

The advantage can easily be illustrated by an example. Empirical growth analysts have tried to show that democracy has a positive effect on growth, without much convincing success, so far.

Especially in poor countries, democratization often leads to instability and frequent changes in power. The resulting uncertainty, in turn, is detrimental for growth. A fairly stable country like Saudi Arabia may fare better economically – even if it is an outright authoritarian system. Such experiences suggest that democracy can only positively influence growth when combined with stability. It took many regressions, unsuccessfully using democracy as an independent variable, until Quinn and Woolley (2004) *replaced* democracy by stability, and suspected that the real correlation lay with stability and growth. In QCA, the relationship between democracy and stability would have been easily visible. The two conditions, democracy and stability, would have appeared combined when the outcome "positive growth process" is present. This, of course, does not exclude that authoritarianism combined with stability could lead to a positive outcome as well. The configurations may be equifinal.

QCA assumes that conditions are embedded in a configuration of conditions, and as such influence the outcome. They are so-called INUS-causes. This acronym stands for **i**nsufficient, but **n**ecessary part of an **u**nnecessary, but sufficient combination of conditions: an "INUS cause is neither individually necessary nor individually sufficient for an outcome." (Mahoney/Goertz 2006: 232) Only the combination of conditions as a whole can be sufficient or necessary.

This leads to the main distinction of sufficient and necessary conditions of an outcome, which are the main concepts of causation used in qualitative analysis. (Mahoney/Goertz 2006: 232; Schneider/Wagemann 2007: 31ff)

A sufficient condition in QCA is a factor A[5] or a combination of factors A*B leading to an outcome Y. It is important that Y is present every time A *and* B are present. Theoretically, there must not be a single case of the conditions A and B being fulfilled, while Y is not. On the other hand, there may be conditions and combinations of conditions other than A and B which equally lead to Y. Assuming only one condition, A, and the outcome Y, the relationship of a sufficient outcome can be displayed as in table 8:

On the contrary, a necessary condition A, or a necessary combination of conditions A*B, is always present when Y is present. There must not be a single case of Y being present while A or respectively A*B are not fulfilled. However, there may be cases of A being fulfilled while Y is not present.

Table 8 Sufficient Conditions in QCA

Condition (A)	Outcome (Y)	
0	0	possible (but possibly not relevant)
0	1	possible (but possibly not relevant)
1	0	not possible (conclusion would be: A is not sufficient for Y!)
1	1	possible and relevant

(*This table has been taken and translated from Schneider/Wagemann 2007: 34. 0 stands for "not present", while 1 means "present".*)

The example of a necessary condition can be illustrated as in table 9.

Table 9 Necessary Conditions in QCA

Condition (A)	Outcome (Y)	
0	0	possible (but possibly not relevant)
0	1	not possible (conclusion would be: A is not necessary for Y!)
1	0	possible (but possibly not relevant)
1	1	possible and relevant

(*This table has been taken and translated from Schneider/Wagemann 2007: 38. 0 stands for "not present", while 1 means "present".*)

While I can conclude from a sufficient condition A that the outcome Y must be given, it is the other way round with a necessary condition. I can conclude from the given outcome Y that A (if it is a necessary condition) must be present. This is symbolized in the formal writing:

$A \rightarrow Y$ (A is a sufficient condition) and $A \leftarrow Y$ (A is a necessary condition). Of course, a sufficient condition might be simultaneously a necessary one and *vice versa*. This would read as $A \leftrightarrow Y$.

Finding out the necessary and sufficient conditions for an outcome is the main procedure of QCA. In this study, this means that those conditions which are necessary for non-oil growth and the configurations which are sufficient for non-oil growth will have to be found out.

The above cited truth tables are simple, but as soon as more than three conditions enter the analysis, truth tables grow out of proportion. Eight conditions, for example, produce $2^8 = 256$ rows. This is why software is used. (Ragin/Davey 2012) While fsQCA can meanwhile be performed by STATA and R, I will use the fsQCA software.

In a first step, every single condition has to be tested for necessity. This regards the condition itself but also its complement, which is the absence of the condition. Both have to be tested for the present and the absent outcome. Depending on the literature, these necessary conditions are then excluded from the analysis of sufficient conditions for the respective outcome (absent or present).

The analysis for sufficient conditions tests at first if any single condition is sufficient taken separately. If this is not the case, a model or combination of conditions can be tested for both the present and the absent outcome. The software yields three different kinds of rows in the truth table, while every row stands for a possible combination.

There are rows with cases, i.e., cases that are members of the entire configuration. Among those, there are consistent rows, and rows which are not consistent. However, there will also be rows without any cases. These rows are called logical remainders. The underlying problem is the limited diversity of reality. Some combinations do not occur in the sample, some may not even occur at all. For the analysis, these rows are deleted, but they are used by the software for the different solutions. The computer usually produces three kinds of "results" for the truth table analysis.

The first one is the complex solution. For an analysis of four conditions A, B, C, and D and the outcome Y, the complex solution could look as follows (the swung dash indicates that the following condition is absent):[6]

Equation 1:

$$\sim A^*B^*\sim C^*\sim D + \sim A^*\sim B^*\sim C^*D + \sim A^*\sim B^*\sim C^*\sim D \rightarrow Y$$
$$\text{(or alternatively aBcd + abcD + abcd} \rightarrow \text{Y)}$$

Without the use of logical remainders, this can be reduced to equation 2:

Equation 2:

$$\sim A^*\sim C^*\sim D + \sim A^*\sim B^*\sim C \rightarrow Y$$

At this point, this solution cannot be shortened any more without the use of the logical remainders.

If the computer makes assumptions about the logical remainders, it can reduce these solutions to equation 3:

Equation 3:

$$\sim C \rightarrow Y$$

However, the software makes these assumptions mathematically and cannot judge the plausibility of its assumptions. In some cases, it makes assumptions which contradict earlier findings: condition A, for example, may have been identified as necessary for ~Y, while the computer may link rows of the truth table to this outcome even if this necessary condition is not present. However, Ragin and Sonnett (2004) have developed a third, intermediate solution. This intermediate solution only admits assumptions which are super-sets of the complex solution. In the above-mentioned example, this could look as follows:

Equation 4:

$$\sim A^* \sim C + \sim D \rightarrow Y$$

In this study, I will mainly rely on the intermediate and complex solutions. If there are both necessary and sufficient conditions, these will be given separately as the final results.

So far, I have used crisp sets which means that a condition or an outcome is either present or not. The conditions have to be coded as either 0 (not present) or 1 (present). In social sciences, the researcher is seldom confronted with such clear cases. Consider the case of "democracy". Comparing Norway to Saudi Arabia, one may agree that Norway receives a "1" for "democracy present" and Saudi Arabia "0" for "democracy not present". The differences between Saudi Arabia and Kuwait, however, are much less clear. Kuwait is certainly less democratic than Norway but probably more democratic than Saudi Arabia.

Of course, every case is unique, but for systematic comparisons, certain categories are necessary. Fuzzy sets can be understood as a compromise between the individual diversity of cases and the need for quantification. In fsQCA, each condition can be seen as a set with different degrees of membership. Qualitatively, country X can belong to the group of democracies, but quantitatively, it may belong less to this group than Norway. In numbers, this means that there are not only two values, 0 (non democratic) and 1 (democratic), but also values in between such as 0.2, 0.6 or 0.8. Note that 0.5 means that a case is truly in between: it is neither democratic nor undemocratic and it is impossible to put it into one of the two qualitative categories. Methodologically, these cases are problematic and should be avoided. A fuzzy value of 0.5 makes it impossible to assign a case to only one configuration of conditions.

It is also interesting that each case has simultaneous membership in a condition *and* in its opposite. Indonesia, for example, might be a 0.8 member in the group of democratic countries and this automatically makes it a 0.2 member in the group of undemocratic countries.

The question is how to assign fuzzy values to the cases. In most cases, this study works with interval-scaled data, which allows for adopting also more fuzzy values between the main mile stones. (Ragin 2008) The software provides a calibration function to this end, but nevertheless, it is important to define at least three values: when is a full membership reached; when is a case definitively not a member; and at what point is the division for countries that are in the group from countries which are out of the group? These questions have to be answered based on theoretical reasoning even if mean or median values sometimes help to make a decision. It is also important to look at the nature of the raw data: the concept of democracy and its measurement differs from numeric variables such as GDP *per capita*. It is also important to note that it makes a big difference if a condition is assessed in relation to the countries of the sample only or to the entire world.

For concepts like democracy, it may seem to make little sense to use complex fuzzy values. How should the fuzzy value 0.87 be theoretically justified? And what distinguishes 0.87 from 0.86? (Ragin 2000; Schneider/Wagemann 2007: 173-185) In this study, however, fuzzy values with two decimal places will be used, because the raw data are average values of more than 30 years. During such a long period of time, countries may have passed from dictatorships to imperfect democracies or *vice versa,* and small differences of fuzzy values reflect the duration of a phenomenon. Nevertheless, this problem points to a flaw of QCA. Time aspects, and in general the aspect of developing conditions, are still neglected by the basic method.[7] (de Meur/Rihoux/Yamasaki 2008: 161-162)

The logical operators of Boolean algebra can also be applied to fuzzy sets but principles are less obvious than with crisp sets. Assume, for example, two cases and three conditions, A, B, and C and the following data (table 10) for two cases:

Table 10 **Fuzzy Sets and Membership (I)**

	A	**B**	**C**	**Outcome Y**
Case 1	0.8	0.9	1.0	0.8
Case 2	0.2	0.4	0.3	0.2

If I want to know the degree of membership of case 1 in A*B*C (logical "and", corresponding to the intersection of A, B, and C, sign: *), I have to consider the smallest value, which is 0.8. The membership degree of case 1 in A*B*C is restricted by the lowest value. Case 2 is a little more complicated. Since all values are below 0.5, the ideal configuration case 2 comes closest to is ~A*~B*~C. Hence, I have to take the complement values (table 11), the values case 1 has in ~A, ~B and ~C.

Table 11 **Fuzzy Sets and Membership (II)**

	~A	**~B**	**~C**	**Outcome ~Y**
Case 2	0.8	0.6	0.7	0.8

Case 2 therefore has a membership degree of 0.6 in the ideal constellation it comes closest to: ~A*~B*~C. For any other constellation, the membership degree would be even lower.

The logical "or" (sign: +) works similarly but the membership degree corresponds to the *highest* value. Case 1 has a membership degree of 1.0 in A+B+C, while case 2 has a membership degree of 0.8 in ~A+~B+~C.

Unfortunately, results of QCA or fsQCA are seldom valid for 100 percent of the cases. Among ten cases, there may be a potentially necessary condition A*~B for a certain outcome Y which is valid in 9 of the 10 cases. This means that, in 90 percent of the sample cases,

A*~B is necessary for the outcome Y. In the tenth case A*~B is absent although Y is present. Should the analyst reject A*~B as a necessary condition?

There is no clear answer to this question. The deviating case(s) may be explicable. Their data basis may be exceptionally unreliable and the sample size is big enough to support the argument of a necessary/sufficient condition sufficiently even if a few cases deviate.

Those problems are the rule than an exception, and therefore two quality measures have been introduced that might help to decide: consistency and coverage. (Schneider/ Wagemann 2007: 202-219) In csQCA, it is only interesting to know *if* some cases contradict the assumption of a necessary or sufficient condition. In fsQCA, however, it is also interesting to know to *what extent* they deviate.

Consistency for Sufficient Conditions

Assume a potentially sufficient condition A, which is investments, and an outcome Y, which is growth. If A really is a sufficient condition, the fuzzy value of A must be equal or smaller than the corresponding fuzzy value of Y. ($A_i \leq Y_i$)

However, in several cases, A is bigger than Y, which means that here, investments might have been high but this did not fully translate into a higher growth rate. The role of A as sufficient for Y is put into question. Computing the consistency measure of a sufficient condition helps to decide whether A has to be rejected as a sufficient condition on this ground.

The following formula computes the consistency of a sufficient condition:

Equation 5:

$$\frac{\sum_{i=1}^{I} \min(A_i; Y_i)}{\sum_{i=1}^{I} A_i}$$

It should be 1 if, in all cases, A is smaller or equal to Y, because in all cases, the $\min(A_i, Y_i)$ would be equal to the fuzzy value of A_i. As a consequence, denominator and numerator result in the same value.

If A is in some cases higher than Y, the minimum is the fuzzy value of Y, leading to a numerator which is smaller than the denominator. The consistency is then lower than 1. And it becomes lower the more values of Y are below A and the wider the gap between the two.

If the consistency of the sufficient condition A is high enough (and only then), it makes sense to ask how important A as a path leading to Y is. There are usually different sufficient conditions and one solution might explain more cases than another. This is what Ragin means by "coverage". The coverage of a sufficient condition can be computed with the following formula:

Equation 6:

$$\frac{\sum_{i=1}^{I} \min(A_i; Y_i)}{\sum_{i=1}^{I} Y_i}$$

Here, the numerator becomes lower than the denominator if A is smaller than Y. This means that the high presence of Y is not or not entirely explained by A in those cases. As a result, the coverage is lower.[8]

It is also possible to compute the coverage of all sufficient paths to Y, and for one path only (if there are overlaps of several paths in explaining one case).

As for the necessary conditions, the same formulas but *vice versa* is used. The formula used to compute the consistency of a sufficient condition is used to compute the coverage of a necessary condition. And the formula used to compute the coverage of a sufficient condition can be applied to generate the consistency of a necessary condition.

4.2 Case, Time, and Variable Selection

FsQCA is a "case-oriented" method and case selection plays and important role. In this study, different samples will be used, mainly to allow for methodological triangulation, but also because of the research question. For the regression, which is mainly used for methodological comparison, all countries covered by the data will be used. The sample size reaches from 109 to 183, depending on the variables included. No case is deliberately excluded.

Two main fsQCA analyses will be made. The first sample will consist of the GCC states only. The focus here is on relatively minor differences, as the main feature – oil and gas abundance – is shared by all of them. As will be shown in the analysis, there seems to be a "turning point" between enormously resource rich countries, and those which are resource rich compared to the rest of the world but which cannot expect to live on their resources.

The fact that resource wealth is relative calls for a further sample of countries. The GCC will be analyzed within a larger sample of resource rich countries, which consists of *relatively* resource rich and *enormously* resource rich countries, but not of resource poor countries. The commonality of all are considerable oil reserves, while only one country (Botswana) is added because of its wealth in diamonds and its remarkable path to development as a rentier state. (Beck 2007) This sample contains 25 countries from several continents.[9]

The time frame of the analysis will be from 1973 to 2006, while data will be averaged over this period of time. As for the regression, a panel data analysis had to be rejected because of visibly poor data quality. The same reason stands against splitting up the period of 33 years into several sub-periods for the fsQCA analysis. Especially for the time until 1985, data quality is extremely poor, while data for some conditions is absent before 1996. Although the time frame is rather large, it is also hoped that this levels out data unreliable single data points.

The following subsection will briefly summarize the variables/conditions that have been chosen.

4.3 Operationalization of the Variables

The operationalization of a variable is one of the most important steps of any empirical analysis. Differences in operationalization can change outcomes sensibly, and single variables can differ strongly according to their operators. Problems with operationalization mainly stem from two different sources: the relationship between operators and the variable, and data problems. As in most empirical studies, variables and conditions make uses of a vast amount of data. A list of variables and conditions and their main sources can be found in the Appendix B.

Non-Oil Growth

Operationalizing non-oil[10] growth, many compromises had to be made. On the one hand, there are, ambitious methods, while there is defective data on the other. Furthermore, for the sake of comparability, picking the special data available for oil-exporters caused problems when the same data was needed for resource-poor countries. The resulting problems can be subsumed under the three following topics:

1. theoretical questions
2. special regional circumstances (foreign labor)
3. data quality and data availability

Initially, there is the theoretical question of how to compute non-oil GDP and non-oil growth.

Several approaches have been undertaken so far. Some are very ambitious in trying to reach preciseness, aiming at a better accounting practice in the future. These approaches are usually feasible for cases studies but not for large samples that also include countries without sizeable rentier income. Other methods are very simple and aim at estimating non-oil growth very broadly. Even if data availability rather than theoretical deliberations were decisive for this study, I will shortly explain several methods.

In "Accounting for 'Wasting Asets': Income Measurement for Oil and Mineral-Exporting Rentier States", Stauffer and Lennox actually proposed two methods: one for calculating a non-depletable GDP and another one for calculating an even lower oil- or rent-independent GDP. (Stauffer/Lennox 1984)

The non-depletable GDP is calculated by deducing the rent obtained from natural resources from the GDP. Stauffer and Lennox propose a sophisticated way for the per unit rent calculation. While the minuend is the resource price, there are several subtrahends: the production costs, the capital depreciation and the opportunity cost of the capital (assumed as a real 10 percent rate of return). (Stauffer/Lennox 1984: 14) This is certainly the most precise method to measure resource rents. However, it is not feasible when more than thirty years and more than one hundred countries are involved. It is therefore not astonishing that this method has – to my knowledge – not been applied in any empirical study on resource-abundant economies.

Eric Neumayer (2004: 1636ff), who claims to be the first to look into the resource curse using a rent-adjusted GDP, names three methods. The first is the so called "Hotelling rent"[11]

Equation 7:

$$(P-MC)*R$$

(*Where P is the world market price for the resource, MC the marginal cost of producing it, and R the quantity produced*).

Marginal cost data, however, is not readily available. In most cases, estimated values for average costs are available and this is why the per-unit rent is usually calculated as price per unit minus average production cost per unit (AC) as an alternative method.

Equation 8:

$$(P-AC)*R$$

(*For the rent calculation, P is the price, AC is average costs, and R is the quantity of resources produced.*)

This is the so called "World Bank method". It is probably the most widespread method because the most comprehensive data-sets on resource rents are provided by the World Bank.

Entirely different are rent calculations based on scarcity value concepts. These methods use estimations of the overall amount of the resource stock, the overall value of the remaining resources, or the number of years until the depletion of the resource. Quite naturally, there are only estimates on how much of a resource a country really has, since there are continuing new discoveries; even for the discoveries that have already been made, there are still only estimates. Unforeseeable price developments and technological developments add further uncertainty. Using resource stock estimates and future price estimates multiplies the risk of total miscalculation. Both Neumayer and Stauffer/Lennox therefore reject these methods.

Other options to measure diversification, although not necessarily non-oil growth, are diversification indices such as the Herfindahl-Hirschman index or the Finger–Kreinin measure. Both are available on UNCTADSTAT, starting from 1995. While the Herfindahl-Hirschman measures market concentration, the Finger-Kreinin-Index measures export concentration. Both are relative measures and compare countries to the world average of diversification. Although a lower Herfindahl-Hirschman index (less concentrated market) is mostly a good sign, it can result from a shrinking oil market due to lower prices and does not necessarily mean that the non-oil sector has grown.

Neumayer, who faced the practical problems of time series data in a huge sample more directly than Stauffer and Lennox, finally made the decision to use the "World Bank Method" as will I in this study.

Admittedly, the Stauffer/Lennox method comes closer to the "real rent" because of the capital depreciation and the opportunity cost of capital. I thus have to keep in mind that the rent resulting from the equation *(price – average cost)*quantity* is partly overstated,[12] while its inflation effect on the rest of the economy is understated.

Data for non-oil growth is taken from two sources: oil, gas, metal, forestry, coal, and lignite rents are from the World Bank's Green accounting program. This program used to provide several Excel-sheets containing the data used for the calculation of net adjusted savings contain oil, gas, coal, metal, and forestry rents. Even if oil and gas are the most important sources of rents in the GCC states, all rents from natural resources have been included. As for oil, the World Bank data used the average current (nominal) prices of four oil types per year.

As mentioned above, the rents taken from the World Bank database on Green Accounting are in current prices. It was thus useful to also work with GDP in current prices. Here, *per capita* nominal GDP numbers have been taken from PENN World table 6.3 (Heston/Summers/ Aten 2009)[13]. The population numbers of PENN World Table have been used to compute a *per capita* natural resource rent, which was then subtracted from the *per capita* GDP.

Not only resource rents put normal accounting practices into question. The GCC states are also extreme examples for two phenomena: foreign labor, and investments abroad. Exact numbers for the migrant workers are hard to obtain. Furthermore, the number of migrants can change fast, for example, in times of economic crisis as the Dubai crisis in 2008 has shown. Accordingly, per capita figures are not very reliable because there are only estimates of the local population. More than elsewhere, the sample countries have to be regarded as local entities and not as the sum of their nationals. Furthermore, part of the generated GDP leaves the country in the form of labor remittances. (Adams 2006)

Secondly, governments and private investors invest largely abroad. Subtracting government spending and private consumption from GDP therefore yields the sum of investments abroad and locally, while numbers can hardly be split up into these two components. All GCC governments now have sovereign wealth funds. In some cases, investments are opaque, while only a few prominent positions have become known due to spectacular deals. It seems that accounting for these two phenomena is only possible in a qualitative study.

Data Quality

Computing non-oil GDP, one big problem occurred: in several countries – all of them Arab Gulf states - non-oil GDP turned negative, which means the oil rent alone was higher than the official GDP. The problematic years have seen years of tremendous increases of oil revenues. There can only be speculations on how this happened, and the lack of a clear separation between state income and the income of the royal families or dictaros may provide one explanation.[14] Without solid explanations, however, no corrections of this phenomenon were possible. To avoid the danger of data embellishing, I therefore decided to stick with the "common" growth rates from the PENN World table for these seven countries in the time period from 1973-1981.[15] In most other oil exporting countries, the "common" mean growth rates from 1973 to 1981 only differ slightly from the mean non-oil growth rates. Using the "common" growth rate, it might be possible to come closer to the non-oil growth rate than by manipulating the amount of the rent.

Calculation

The remaining non-oil GDP has been chained by the US-$ GDP deflator, with the year 2000 as a base year. Based on the obtained non-oil GDP in constant prices, growth rates have been calculated. Finally, the average growth rate for the entire period has been used. These growth rates of the twenty-five sample countries have been compared to other studies (Amzugear 2001; Koren/Tenreyro 2010), investment guides, IMF papers (Fasano/Iqbal 2003), and diversification indicators.

For the regression analysis, the variable NDGROWTH consists of these average growth rates. For the fsQCA analyses, fuzzy values had to be assigned to each of the countries. If this part of the analysis only dealt with the six GCC states, this could have been the next step, but the sample of ternty-five is very heterogeneous. Starting from a very high non-depletable GDP means growing a lot to reach a modest growth rate of 2 percent. In theory, poor countries should reach high growth rates more easily. At least this is what convergence theory predicts, and this analysis will not uncritically accept this prediction by "weighting" growth rates according to this seldom-reached ideal.

I thus turned to the second, empirical possibility of judging a country's growth performance. The initial *per capita* GDPs of the sample countries are quite diverse. In real (2000) US $, purchasing power parity and rent adjusted, they ranged from less than US $1000 *per capita* (Nigeria) to around US $40 000 *per capita* (Qatar) in 1973. The sample countries cover the whole range of GDPs in 1973, from the poorest to the richest countries. At this point, the question becomes: what is the mean growth rate a country, having a certain initial GDP, could have reached if it performed very well, acceptably well, not very well, or badly? This question turned my attention to non-sample countries with similar initial GDPs.

I have computed the non-oil *per capita* growth rates of all countries in the world (as long as data were available), as well as their mean growth rates from 1973 to 2006. These countries were then divided into six categories according to their initial non-oil GDP as of 1973. Thus I obtained a scope of growth rates that apparently was possible from a similar starting point. To give just one example: China started with a *per capita* GDP of around US $500 in 1973 and in the following years had mean growth rates of 6 percent. For the category of countries with very low initial GDPs, it would thus have been possible to reach growth rates of 6 percent under "ideal" circumstances. Within the same category, however, other countries had negative growth rates.

In another category, countries with the highest initial non-oil GDP in 1973, such as Luxembourg, Switzerland or the US, mean growth rates from 1974-1985 in all cases were below 1 percent. Not a single country in this category had a higher mean growth rate. If 6 percent is a possible average growth rate in China, 6 percent would have been impossible for countries with high initial GDPs. The overview of possible (table 12) growth rates in the six groups visualizes the scope of what would have been possible. Within each group, I have identified very good performances, fairly good performances, modest performances, and bad performances – each with a certain range of growth rates. As a fixed point, growth performances of 0 and below were always bad performances (fuzzy value 0), while for very poor countries 0 was only a bad performance. Fuzzy value 1 (very good performance) was given to the range of the three highest growth rates of one group in order to avoid judging the group's performance by one extreme outlier. The growth rates corresponding to the other fuzzy values were then calculated proportionally.

This method, however, has one flaw: the GDP of the Gulf states is overrated by income from oil, even if oil rents have been subtracted (see the next section). The use of initial GDP is meant to account for levels of development, which it does not in this case. The stage of development of Qatar in 1973 was hardly the same as in the Netherlands, even if both countries are in the same group. On the other hand, however, the high GDP of Qatar is the basis for the growth rate calculation, and here the mathematical concept of the percentage takes effect. Be it due to the rentier multiplier or not, higher numbers need to grow more to reach the same percentage as a lower number. As a compromise, Brunei, Kuwait, Qatar, and Saudi Arabia have been downgraded to group 3 (see table 12, bold). This group also includes Bahrain, which probably had the highest level of development among the GCC states at that time. Oman remains in group 2, because development started later.

Table 12 **Groups by Initial GDP as of 1973**

Year of Initial GDP	Countries and Categories
1973	Group 1 (<$2000): Nigeria, Botswana, Iraq, Indonesia Group 2 ($2000 – $4000): Algeria, Ecuador, Egypt, Malaysia, Oman Group 3 ($4000 – $7000): Bahrain, Brazil, Iran, Mexico, UAE, Venezuela, **Brunei, Kuwait, Qatar, Saudi Arabia** Group 4 ($7000 – $12 000): Gabon Group 5 ($12,000 – $16,000): Canada, Norway, UK Group 6 (>$16,000): Netherlands, USA

Having in mind the above mentioned restrictions, the outcome is calibrated as follows (table 13):

Table 13 Non-Oil Growth – Variables and Fuzzy Values of the Outcome

	NDGROWTH			Attributes
Country	Mean Growth Rate	Group	Fuzzy Values	
Algeria	1.98	2	0.70	fairly good performance
Bahrain	8.29	3	1	very good performance
Botswana	5.94	1	1	very good performance
Brazil	1.14	3	0.30	rather modest performance
Brunei	-4.93	3	0	bad performance
Canada	2	5	0.8	fairly good performance
Ecuador	0.57	2	0.2	bad performance
Egypt	2.21	2	0.7	fairly good performance
Gabon	1.29	4	0.4	rather modest performance
Indonesia	3.73	1	1	very good performance
Iran	1.41	3	0.3	rather modest performance
Iraq	-7	1	0	bad performance
Kuwait	-2.29	3	0	bad performance
Malaysia	4.06	2	1	very good performance
Mexico	1.01	3	0.3	rather modest performance
Netherlands	1.21	6	0.6	fairly good performance
Nigeria	1.35	1	0.4	rather modest performance
Norway	1.32	5	0.6	fairly good performance
Oman	2.5	2	0.8	fairly good performance
Qatar	0.29	3	0.1	bad performance
Saudi Arabia	-1.09	3	0.00	bad performance
United Arab Emirates	7.09	3	1	very good performance
United Kingdom	2.08	5	0.8	fairly good performance
United States	1.84	6	0.7	fairly good performance
Venezuela	1.01	3	0.3	rather modest performance

Sources: World Bank; PENN World table 6.3.

Initial Non-Oil GDP

Initial non-oil GDP (1973) as used for the calculation of the non-depletable growth rates will enter the regression as LOGNDINITIAL, the natural logarithm of the initial GDP as of 1973 which is the widely used form of this variable. (Neumayer 2004)

Initial GDP will not enter the fsQCA-analyses because it has already been used for the calibration of the outcome and can/must therefore be dropped as a condition. Another reason is the doubtfulness of such a condition in truly resource dependent states. As Michael Herb put it, "[t]here are no statistics that could serve as a measure of development that have not been confounded by oil wealth." (Herb 2005: 301) Even non-oil GDP does not reflect the true level of development in the sense of Solow's theory, because "[t]he effect of oil (or other rents) on the economies of the richer rentier states is transformative, not additive." (Herb 2005: 302)

Looking closer at the initial *per capita* GDPs of 1973 (table 14) and the annual average resource rent *per capita*, it is obvious that they are in most cases very close to each other, even roughly covariant. The only exceptions are the developed countries: Canada, the Netherlands, Norway, the UK and the USA have low *per capita* rents yet nevertheless high GDP (see shadowed area in table 14). Again, it seems that and even a GDP which is corrected for natural resource rents is highly inflated by the rentier multiplier effect. Hence there is another reason to drop initial GDP as a condition in the fsQCA analysis. It is a deceiving measure in resource abundant states.

Looking for an Alternative

It could be argued that there may be alternatives reflecting the level of development better. An attempt was made to use the Human Development Index (see table 15), which not only contains data on GDP, but many more factors of human capital, access to health care and water etc. (UNDP 2011) Here it seems that the combination of a small country size and immense oil wealth accounts for higher HDI scores – the result of welfare and public investment which can be more generous when there are higher rents and fewer people. Today, Bahrain has overtaken Kuwait and ranks just below Qatar. However, the differences between the six countries remain minor. The biggest flaw of such an Index is the fact that it already contains several factors which should enter the analysis as separate factors.

Usually, investments are operationalized as "investment share of GDP" or "gross fixed capital investment". "Investment share of GDP" seems to indicate whether a country saves and invests "enough". Again, the regression analysis will use this common variable as this is an acceptable measure for many but not all countries in the world.

In the case of resource abundant and especially resource dependent countries, this operationalization poses a special problem that can be illustrated by the example in

Table 14 Initial GDP as of 1973 and Rents

Country	Annual Average Resource Rent	Initial Non-Oil GDP as of 1973	RENT (fuzzy set)	NDINITIAL (fuzzy set)
Algeria	512.99	4716	0.04	0.09
Bahrain	3470.91	9257	0.39	0.42
Botswana	874.09	3659.33	0.06	0.06
Brazil	53.68	6344.67	0.03	0.17
Brunei	8689.62	22108.81	0.99	0.97
Canada	15222.93	756	0.05	0.95
Ecuador	208.1	4226	0.03	0.08
Egypt	106.16	3340.67	0.03	0.05
Gabon	1628.44	8633.33	0.11	0.36
Indonesia	71.64	2298.68	0.03	0.04
Iran	494.52	5809.33	0.04	0.14
Iraq	760.13	1558	0.05	0.03
Kuwait	8235.17	14991	0.99	0.82
Malaysia	344.97	6366	0.04	0.17
Mexico	269.24	6633	0.04	0.19
Netherlands	375.99	20109.67	0.04	0.95
Nigeria	148.94	915.33	0.03	0.02
Norway	3508.96	20335.33	0.39	0.96
Oman	2446.9	9581.67	0.2	0.46
Qatar	10596.64	19662.33	1	0.95
Saudi Arabia	3953.18	10486.33	0.49	0.54
United Arab Emirates	9439.3	13263	1	0.73
United Kingdom	334.34	18009.67	0.04	0.92
United States	312.54	25683.67	0.04	0.99
Venezuela	1102.28	5835.67	0.07	0.14

Sources: World Bank; PENN World table 6.3.

Table 15 Human Development Index for the GCC States as of 2007

	HDI 2007
Bahrain	0.895
Kuwait	0.916
Oman	0.846
Qatar	0.91
Saudi Arabia	0.843
UAE	0.903

Source: UNDP 2009.

Investments

table 16. Consider two countries, A and B, both having a *per capita* GDP of US $100. Country A has a non-oil GDP of US $90, while B, being a huge oil-exporter, has a non-oil GDP of US $40. Country A invests 30 percent of its GDP, while country B invests only 20 percent.

Table 16 Investment/GDP Ratios in Resource Abundant Countries

	GDP	Non-Oil GDP	Investment share	Investment *per capita*	Investments/ Non-Oil-GDP
Country A	$100	$90	30%	$30	one third (33.33%)
Country B	$100	$40	30%	$20	one half (50%)

In absolute terms, country A invests US $30, while country B invests only US $20. Assuming that non-oil GDP better reflects the development of an economy, A is much more developed than B. It has also lower income from oil which makes it less likely that the US $90 stems entirely from oil rents via the rentier multiplier effect. A has much more industry, services, and enterprises than B. This means that A has more to invest in than B. For country A, it may be possible to invest US $30, while country B may be able to save 30 percent of its GDP (US $30) as well but the economy is just not developed enough to absorb this sum on a reasonable purpose. So, country B "only" invests US $20, which is already one half of its non-oil GDP. Both countries may have made the best policy decision, but for the time being nobody would be able to judge what the best policy decision would have been. The situation of having more money than can be invested is a unique situation of resource abundant countries and most of them have not made the ideal investments in their country. It is impossible to know whether, to use an extreme example, 100 percent of the existing non-oil GDP could have been invested, or if this would have necessarily led to wrong decisions and wasted investments.

The measure "investments share of GDP" will therefore not be a measure of the quality of investment decisions. The alternative would be to calculate an investment share of non-oil GDP. For resource dependent countries having rather low non-oil GDP, it would be easy to reach a high investment rate. Nevertheless, this measure also has its flaws because investments also include investments into the resource sector, and these investments are among the most costly.

Again, the use of rent-inflated GDP numbers poses a problem to operationalization. For the six GCC states, therefore, I tried to find a measure without any such numbers. I compared the investment *per capita* to the rent *per capita*. This of course does not measure the effect of investments on the economy, but it is rather an evaluation of the investment decisions. The underlying assumption is that rents could always be saved. It could be argued that, again, resource poorer states have less difficulty investing a higher share of their average *per capita* rent income whereas those countries with extremely high rents could only invest small shares of this income. The case of the UAE, where extremely high income from rents poured in, however, shows that very high shares of the rentier income have been invested is the country. This condition (INVESTCAPITA) looks as follows (table 17).

Table 17 Calculation of INVESTCAPITA

Country	Investment *per capita*	Rent *per capita*	Share of Rents Invested (INVESTCAPITA)	Fuzzy Values
Bahrain	1728.73	3270.92	0.53	0.65
Kuwait	3137.43	8235.17	0.38	0.08
Oman	1418.75	2446.9	0.58	0.83
Qatar	4626.15	10596.64	0.44	0.23
Saudi Arabia	1613.54	3953.18	0.41	0.14
UAE	6523.46	9439.3	0.69	0.98

Sources: World Bank, PENN World Table 6.3.

Since such a measurement is unusual, there is no theoretical opinion on how much of the rents pouring in should be reinvested and can be reinvested. Looking at the data, it seems that a turning point could be at 50 percent. A bad investment behavior (0) will be set at 35 percent, while 65 percent is needed to receive the fuzzy value 1 for a good investment behavior.

This condition will only be used for the analysis of the six GCC countries. In the larger sample, this measure is not appropriate for most of the sample countries. Here, the common investment condition (INVEST) is used, bearing in mind that this is a trade-off (see table 18).

Investment share of GDP is taken from the World Bank Data (gross fixed capital investment as a percentage of GDP). The calibration was sample oriented in this case because of the special circumstances in resource abundant countries, where investments do not necessarily presuppose savings. The fuzzy value of 1 was reached with 30 percent investment share of GDP, while 10 percent mark the bottom line (fuzzy value 0). The turning point was set at 20 percent.[16]

Human Capital and Women

Human capital is a standard variable in any growth regression. Nevertheless, there is no agreement on suitable indicators. Widely used are literacy rates, average years of schooling, the number of students enrolled in primary, secondary and tertiary education, or the number of university graduates. In a recent study, Ludger Wößmann proved that assessing the impact of a country's human capital on growth depends largely on the indicator used. (Wößmann 2004) Many indicators are doubtful and suffer from qualitative concerns: not only do the numbers of years at school count, but so does the drastically differing quality of the schools

In many developing countries, there is a huge gap between the quality and *curricula* of expensive private and public schools. Furthermore, university degrees and formations differ with respect to their economic relevance, and several researchers have weighted education by their return in salaries. Most likely, science, engineering and economics will be given more importance by these measures than the humanities. However, these refinements of human capital measures could only be calculated for the most advanced economies. (Wößmann 2004: 28)

Assuming differences in the quality of education, Hanushek and Kimko (2000) compiled a data set on education and labor force quality in 2000. As a critique of standard quantitative human capital measures, they proposed a measure that took into account qualitative differences of schools across different countries using the performance of students in international tests in mathematics and science. (Hanushek/Kimko 2000: 1185) Based on these test results, Hanushek and Kimko constructed two quality measures (QL1 and QL2) and both of them turned out to be highly significant and positively correlated with growth. Still, it should be noted that the sample size is only 30 or 31 countries, and using estimates for further countries, 78 or 80. Wößmann, in turn, provides estimates for quality of schooling for most countries. In table 19, the average

Table 18 Investment Share of GDP (Large Data Sample)

Country	Gross Fix Capital Investment/ GDP	Fuzzy Sets
Algeria	19.07	0.41
Bahrain	26.67	0.88
Botswana	22.77	0.7
Brazil	19.09	0.42
Brunei	26.57	0.88
Canada	29.69	0.95
Ecuador	26.63	0.88
Egypt	8.04	0.01
Gabon	12.12	0.05
Indonesia	17.9	0.31
Iran	33.51	0.98
Iraq	18.54	0.37
Kuwait	22.59	0.69
Malaysia	24.6	0.8
Mexico	20.38	0.53
Netherlands	24	0.77
Nigeria	9.74	0.02
Norway	33.47	0.98
Oman	23.72	0.75
Qatar	26.12	0.86
Saudi Arabia	31.85	0.97
United Arab Emirates	59.08	1
United Kingdom	19.53	0.46
United States	22.04	0.65
Venezuela	22.51	0.68

Sources: World Bank.

years of schooling in the period under scrutiny (taken from the Barro-Lee data-set on educational achievements) can be compared to Wößmann's estimates of schooling quality. The average years of schooling reflect more or less what is known about human capital in the region. I would have expected Bahrain to have a higher figure. Bahrain is known to have had the first schools in the region, and today Bahrain has the highest share of nationals in the private sector in the GCC. This may be due to a less pronounced rentier mentality, but it also seems that a lot has been achieved regarding vocational training. (Alkhozai 2008)

Table 19 Quantitative and Qualitative Measures of Human Capital

	Bahrain	Kuwait	Oman	Qatar	Saudi Arabia	UAE
AVSCHOOLING	4	5.1	4	4	3.4	4
Quality of Schooling	0.226	0.229	0.341	0.444	0.341	0.444

Sources: Wößmann; Barro Lee data set on educational attainment.

The quality of schooling index, however, assigns Bahrain and Kuwait the lowest quality. This may be explainable. It can be expected that the quality of schools and universities is heterogeneous across the GCC states. For example, the UAE and Qatar especially have founded schools, colleges and universities with partnerships of Western elite institutions. Here, Bahrain and Kuwait lag behind. Nevertheless, it is hard to explain how schooling quality in Bahrain and Kuwait should be lower than in Saudi Arabia, where *curricula* are well known to be unsuitable for a modern society and economy even if the number of elite colleges and universities grows. (Prokop 2005)

Due to these data problems, I will not use the quality of schooling index. For the regression, AVSCHOOLING will be used. For the smaller sample of the GCC states, differences between the average years of schooling appear to be too small to be reliable. On the entire scale, where some countries reach values of around 12, the variance between the GCC states is too small and not reliable enough. For the first fsQCA analysis, the variable will thus be dropped.

For the second analysis and the bigger sample, the average years of schooling will be used for the calibration of fuzzy sets. However, this condition (HUMANCAP) will be included under reserve only: There are too many doubts about the accuracy of this operationalization.

Particularly in the GCC states, human capital is distorted by empirically proven low work motivation (Jones 2015), labor migration, and the low participation rates of women in the labor market.

As for labor migration, numerous skilled and unskilled migrant workers work in the GCC states and the effects of "domestic" human capital on growth might be mitigated. The benefit of better education does not always accrue to the country that has invested in education. In the GCC states, effective human capital is certainly much higher than the human capital of the nationals. Secondly, women in the GCC states have come to be well educated nowadays. Their share in all levels of schools and universities has risen significantly in the last decades. This has a very important positive impact on the variable "human capital". Nevertheless, Robert Barro (1997: 13) found that the education of men was highly significant for growth while the education of the female population was not. The reason for this might be with a fact very pronounced in the GCC countries: whereas the share of women in schools and universities has risen to more than 50 percent, their participation in professional life has risen only slowly.

In 1980, in most GCC states less than 15 percent (in Saudi Arabia even less than 8 percent) of the labor force was female. Even if the share of women in the labor force rose constantly in all GCC states, only Kuwait reached a percentage of around 25 percent in 2005. In all other GCC states, the percentage remained below 20 percent. Neglecting for the fact that the education of women has certainly had positive personal effects as well as a positive impact on child education at home, this means that educational efforts fail to have the positive effect on the labor force they could have.

This phenomenon is worth accounting for. As for the GCC states and the small-n analysis, women's labor force participation rates will become part of the condition ISLAM. (see below) It is important to note that there is not necessarily only a connection with Islam. The separation of male and female spaces in public, for example, also raises the need for female staff for female spaces – such as a female security crew at airports. Nonetheless, connecting the two phenomena in one condition helps to avoid the introduction of further conditions at such small a sample size. This is also opportune, because from a theoretical point of view, female labor force participation rates in the GCC states are *all* rather low and would be assigned fuzzy values lower than 0.5. The result would be a lack of variance in the small sample. For the regression and for the bigger QCA sample a variable/condition named WOMENLAB/WOMENLAB will be introduced.

For the calibration, I assume Norway as an "ideal" case, where 70 percent of women between 15 and 64 have a job. The Scandinavian countries are usually considered role models when it comes to gender issues in the labor market and day care for children. The turning point is at 40 percent, while 15 percent or fewer qualify for a fuzzy value of 0. Interestingly, African countries such as Gabon and Botswana also reach a very high percentage. None of the GCC states reaches a fuzzy value of more than 0.5. Looking at the percentage of women in the total labor force in table 20, the GCC states perform even worse. The overlap of male migrant workers may add to this picture. However, it should be noted that data sources are unclear about the influence of labor migration on these numbers, and also about what qualifies as "labor force participation" (for example, in African countries). Numbers may refer to the entire population or only to the nationals.

Hence, this variable/condition should be interpreted with care. Due to these doubts, I will also test fertility rates as a variable reflecting gender issues.

Table 20 Women in the Labor Force

Country	Female Labor Force Participation Rate (of Women between 15 and 64) – 1980-2005	Fuzzy Values	Percentage of Women in Total Labor Force (1980-2005)
Algeria	26.31	0.06	24.34
Bahrain	26.4	0.06	16.12
Botswana	57.11	0.93	44.27
Brazil	51.05	0.84	37.45
Brunei	42.23	0.58	31.16
Canada	65.9	0.98	43.98
Ecuador	40.85	0.53	31.93
Egypt	21.95	0.03	22.79
Gabon	63.99	0.97	43.51
Indonesia	49.7	0.81	37.32
Iran	26.83	0.07	24.09
Iraq	17.38	0.01	17.24
Kuwait	35.49	0.29	20.54
Malaysia	44.92	0.68	34.93
Mexico	37.08	0.36	31.55
Netherlands	56.14	0.92	40.16
Nigeria	48.47	0.78	35.75
Norway	70.18	1	45.03
Oman	17.94	0.03	12.36
Qatar	29.73	0.11	11.76
Saudi Arabia	14.64	0	11.65
United Arab Emirates	27.29	0.03	10.27
United Kingdom	65.52	0.98	43.88
United States	66.26	0.98	44.5
Venezuela	42.81	0.6	33.16

Source: World Bank.

Population Growth and Fertility Rates

Closely linked to female participation in the labor force are fertility rates. As a rule, it seems that lower birth rates correlate with a higher participation in the labor force. Firstly however, this is not always an indirectly-proportional relationship: There may be gradual differences depending on the availability of kindergartens and nurseries and a country may have higher birth rates *and* higher female labor force participation rates. Secondly, African societies seem to be exceptional. Though usually not in employment as understood in developed countries, women participate in the labor force because they are the breadwinners of their families. In the three African states in the sample, high birth rates go hand in hand with high female labor force participation. The fertility rates can be taken from table 21.

Calibration used the following thresholds: full membership in the "high fertility" countries is attained at 4 births per women. At such a birth rate, the population theoretically doubles within one generation. Full membership in low fertility countries – considered positive from the theoretical point of view – is attained at 2 births per women.[17] At such a pace, the society theoretically reproduces itself and does not decrease. At lower rates, fears of decreasing manpower usually arise. The turning point was set at 3 births per women. The result is a polarized sample: fuzzy sets are either very low or very high. Few countries have fuzzy values in between. This rather strict calibration was adopted because high population growth has amultiplier effect on problems. It decreases *per capita* growth, not only because the population rises but also because the population becomes, at least in the short run, younger. For the generation to come, this means spending without productive activity. The labor force further decreases when women stay at home for child rearing. And the effects are probably worse if children do not get the education they need for the labor market.

Resource Abundance, Resource Dependence, Resource Reserves, and Rentier State Theory

According to the theory of Dutch disease or the resource curse, negative effects on growth increase with the amount of rents. Most importantly, resource windfalls, which can happen in any country, are suspected to trigger Dutch disease. Nevertheless, when it comes to the empirical tests of Dutch disease, the resource variable is often a proxy for the resource dependence of an economy, reflected in the share of revenues from resource exports in GDP or in total exports. Little attention has been paid to this important difference thus far. Instead, notions such as resource-abundant, resource-intense, resource-rich, sometimes even resource-dependent have been interchanged thoughtlessly. (Dietz/Neumayer/de Soysa 2007; Neumayer 2004; Sachs/Warner 1995/97) In this study, measures of resource dependence shall be used distinctly from resource abundance measures.

Measures of resource dependence set resource rents or exports in relation to GDP or total exports. If the share of natural resource rents or exports in GDP is very high, a country is

Table 21 Births per Woman and Fuzzy Values

Country	Birth per Woman (averaged from 1980 to 2006)	Fuzzy Values FERTILITY
Algeria	4.57	0.99
Bahrain	3.63	0.87
Botswana	4.54	0.99
Brazil	3.06	0.54
Brunei	3.23	0.67
Canada	1.65	0.02
Ecuador	3.81	0.92
Egypt	4.25	0.98
Gabon	4.87	1
Indonesia	3.26	0.69
Iran	4.39	0.98
Iraq	5.86	1
Kuwait	3.75	0.9
Malaysia	3.6	0.86
Mexico	3.47	0.8
Netherlands	1.63	0.02
Nigeria	6.41	1
Norway	1.83	0.03
Oman	5.75	1
Qatar	4.33	0.98
Saudi Arabia	5.66	1
United Arab Emirates	4.02	0.96
United Kingdom	1.77	0.02
United States	1.94	0.04
Venezuela	3.45	0.79

Source: World Bank.

dependent on natural resources. The same is true if almost all exports consist of natural resources. High natural resource rent/GDP ratios can stem from two situations: either a country produces enormous amounts of natural resources or it produces little or nothing other than natural resources. If these two situations are combined, rent/GDP ratios can reach more than 90 percent.

As another measure of natural resource dependence, even if of more interest to political scientists, "rentierism" should be mentioned. This concept is based on the share of oil revenues in total government revenues. As mentioned in the section on rentier state theory, Luciani (Luciani/Beblawi 1987: 13) would speak of a rentier (or "allocation") state once more than 40 percent of government revenues stem from rents. High "rentier" values can equally result from two situations: a country has either massive resource rents, or it levies little or no taxes. The combination of both can boost "rentier" percentages to about 90 percent. (Luciani 1987: 71)

Most natural resource variables are proxies for resource dependence. An example is SXP, the variable used by Sachs and Warner, which is the percentage of primary product exports of the GNP as of 1970. A variable on "rentierism" was used by Michael Herb. (Herb 2002) I took the latter and averaged it (RENTIER) from Herb's data set. Both variables contain more information than just resource abundance.

First of all, the share of primary product *exports* means that only those resources are included that are not further used by a country in its manufacturing sector or in order to produce energy. An industrialized country with a big manufacturing sector and a lot of energy consumption will export a smaller share of its natural resource production than an equally resource-rich, non industrialized country. Using *exports* as a measure will, as a consequence, underrate the resource abundance of industrialized countries. Second, developing countries, having lower GDPs or GNPs, will automatically have higher percentages of natural resource exports over GNP. They may even export less oil or other resources than an equally big richer country, but their resource "abundance" as measured by SXP will be the same or even higher. Similarly resource *abundant* countries such as Norway and Oman are assigned entirely different values by SXP. While both had an average annual *per capita* resource rent of around US $2500 from 1973 to 2006, Norway is assigned a 0.14 and Oman a 0.89. This is obviously due to Norway's higher GNP and its degree of industrialization. Compared to high non-oil GDP or GNP, resource abundance appears to be smaller if resource dependence measures are used. In the case of Oman, on the other hand, there was only a very small non-oil economy in 1970, which makes the resource sector *appear* larger. In terms of resource abundance, however, the two countries are comparable.

About the same is true for rents in government income (RENTIER). The degree of "rentierism" will be higher if a government does not dispose of any other sources of income, especially if there is very little economic activity to tax (or no will to tax it).

From this reasoning, it becomes evident that SXP and RENTIER are biased. The resource wealth of poor, non-industrialized countries is overstated, while the resource abundance of industrialized, rich countries is underrated. The resource curse can thus become a self-fulfilling prophecy because the variables SXP and RENTIER do not indicate if a country *is* resource rich, only if its economy or government *depends* on resources.[18] However, the puzzle of the resource curse is the astonishing fact that an *abundance* of resources – considered to be a blessing by nature – does not apparently contribute to the growth performance of the countries possessing them. It does not ask why resource *dependent* economies grow slowly.

For Dutch disease, it is a boom in the resource sector that is supposed to cause the resource movement or the spending effect; it is not created by an already-existing dependence of an economy on its resources. The Netherlands, the country that lent its name to Dutch disease, was certainly not resource-dependent when natural gas was discovered.

Because of these considerations, I will test four different measures of resource abundance and dependence. Raw data for these variables can be taken from table 22. As for resource abundance, two "concepts" will be tested by two variables/conditions for each of them: average natural resources rents *per capita* from 1973 to 2006 result in the variable/condition RENT/rent. RENT/rent indicates the amount of resource rents that really pour into a country.[19] The calculation follows the rent calculation which was effectuated for the non-oil GDP. The underlying theory may be Dutch disease and any theoretical environment relying on rent-seeking or rentier mentality.

To reflect the importance of incentives more clearly, the variable/condition RESERVES/reserves will be introduced. RESERVES/reserves indicate to what extent a country can expect to live from its resource rents in the future, although this depends on several uncertainties such as future demand, prices, or substitutes. Unfortunately, only recent data was available for the amount of natural resource reserves of a country, and data does not cover the same variety of resources as the RENT variable. Included are oil, gas, coal, silver, gold, bauxite, iron ore, and zinc. The bigger problem lay with the combination of different resources. As future prices are unknown, resources could not just be added up to a sum of US $ and divided by the population. The most problematic case was Botswana with future diamond production and diamond "reserves" being extremely uncertain. Instead, they were assigned separate fuzzy values based on the comparison of the sample countries for each resource. This proceeding was also used for the variable RESERVES. RESERVES/reserves are meant to be rough indicators for a potential psychological effect and a vague perception of how long the resource wealth could sustain the economy. Additionally, I will test SXP/sxp as well as RENTIER/rentieri as a proxy for resource *dependence*.

Again, the question of fuzzy set calibration arises. As for sxp, the condition varies between 0 and 1 and the complexity of factors entering this condition/variable makes it difficult to weight.

Table 22 **Raw Data for RENT, RESERVES, SXP and RENTIERI and II**

Country	Annual *per capita* Resource Rents, averaged from 1973 to 2006	Natural Gas Reserves in m³/capita	Oil Reserves in bbl/capita	Sxp as of 1970	Rents as a Percentage of Government Revenue
Algeria	512.99	139132.82	377.04	0.19	0.53
Bahrain	3470.91	186796.57***	**263.87*****	0.4	0.59
Botswana	874.09	0	0	0.05**	0.64**
Brazil	53.68	1984.64	68.62	0.05	0
Brunei	8689.62	1068673.48	3008.04	0.79	0.87
Canada	756.37	51267.62	5567.54	0.1	0
Ecuador	208.1	683.97	357.36	0.11	0.47
Egypt	106.16	22796.66	50.93	0.07	0.12
Gabon	1628.44	20787.9	1468.07	0.33	0.5
Indonesia	71.64	13792.15	18.34	0.11	0.49
Iran	494.52	417142.44	2023.32	0.12	0.55
Iraq	760.13	130131.36	4720.85	0.3	0.55
Kuwait	8235.17	1281428.57***	**71 535.71*****	0.73	0.88
Malaysia	344.97	94398.4	160.68	0.37	0.23
Mexico	269.24	3652.14	102.89	0.02	0.23
Netherlands	375.99	82669.33	6.14	0.15	0.13
Nigeria	148.94	40517.78	271.41	0.14	0.71
Norway	3508.96	503707.58	1454.72	0.1	0.09
Oman	2446.9	482670.45***	**3 125*****	0.89	0.81
Qatar	10596.64	132947368.42***	**80 052.63*****	0.8*	0.87
Saudi Arabia	3953.18	332304.77***	**12 109.44*****	0.6	0.8
United Arab Emirates	9439.3	7340609.86***	**118 252.62*****	0.68	0.84
United Kingdom	334.34	5730.83	56.99	0.03	0
United States	312.54	22921.42	72.6	0.01	0
Venezuela	1102.28	185248.98	3803.73	0.24	0.57

Sources: World Bank, BP, Oil & Gas Journal; Sachs&Warner data set; Michael Herb data set;

*SXP for Qatar is my own estimation.

**Botswana: difference between SXP (1970) and RENTIER is explained by the fact that diamond mining only started in 1973. Today diamonds account for 30 percent of Botswana's GDP and for 70 percent of the export earnings. (IRIN 2009)

***In the case of the GCC states, reserves *per capita* were calculated with estimates of the local population.

When deciding over the fuzzy set calibration (see table 23), it is important to keep in mind that *all* sample-countries have substantial revenue from resources. If they were compared to all countries of the world, probably all of them would be more or less "resource rich", given that the median yearly *per capita* rent from resources of all countries in the world is around US $15, while in this sample of 25 countries, the resource "poorest" country received an yearly average of around US $53 *per capita* (Brazil). The question is not "who is resource rich and who not?", but rather "who is more resource rich than the others?" The turning point for RENT was set at US $4000 *per capita* as a yearly average, while the fuzzy value 1 was reached at US $9000. A "0" membership score would have been reached at US $15, corresponding to "below the median natural resource rent".

For the condition RESERVES, another difficulty arose: The basis is *per capita* oil and gas reserves, which can hardly be added. Here, single fuzzy sets for oil and gas reserves were estimated and averaged.

As for the share of resource rents in government income, two different calibrations will be used. The condition RENTIERI will only be used for the fsQCA analysis of the larger sample. Here, the turning point will be set at 55 percent. At 10 percent and below, the fuzzy value 0 was assigned. The fuzzy value of 1 was reached at 90 percent of natural resource rents in government income. Another calibration, and as a result another condition, RENTIERII will be explained in chapter five.

Islam and Arab Society

As mentioned in the theoretical chapter above, a very careful and balanced approach is needed to construct a condition on Islam. For the smaller sample (GCC states only), a condition based on institutional factors can be constructed. I propose a combination of six factors, adding up to the assessment of the condition ISLAM. These five factors include

1. the general ban on interest as an institution – even if reality contradicts it – that puts into question ongoing banking practice
2. the *obligatory* alms tax
3. the degree of separation of men and women in public (as a cause for additional costs)
4. strict public morals as an impediment to the development of tourism and as a deterrent for foreign investors and experts
5. Islamic law and strong influence on commercial law, raising insecurity levels for foreign investors
6. a very strong threat of local terrorist attacks using Islam as a pretext (if more than in any other country)

From a 0 for "no economically relevant Islamic institutions", each point found to be present will add 0.2 to the score. If the factor is present but only to a modest degree, only 0.1 will be added.

This will add up to the following picture in table 24:

Table 23 Fuzzy Values for Natural Resource Conditions

Country	RENT	RESERVEs	SXP	RENTIERI
Algeria	0.04	0.3	0.04	0.46
Bahrain	0.39	0.23	0.18	0.62
Botswana	0.06	0.08	0.02	0.75
Brazil	0.03	0.2	0.02	0.01
Brunei	0.99	0.8	0.95	0.98
Canada	0.05	0.67	0.02	0.01
Ecuador	0.03	0.13	0.02	0.33
Egypt	0.03	0.09	0.02	0.02
Gabon	0.11	0.33	0.12	0.39
Indonesia	0.03	0.18	0.02	0.37
Iran	0.11	0.73	0.03	0.55
Iraq	0.03	0.4	0.1	0.6
Kuwait	0.99	0.97	0.88	0.98
Malaysia	0.04	0.17	0.15	0.06
Mexico	0.04	0.09	0.01	0.06
Netherlands	0.04	0.13	0.03	0.03
Nigeria	0.03	0.14	0.03	0.87
Norway	0.39	0.68	0.02	0.02
Oman	0.2	0.4	0.99	0.96
Qatar	1	1	0.95	0.98
Saudi Arabia	0.49	0.43	0.6	0.95
United Arab Emirates	1	0.9	0.9	0.97
United Kingdom	0.04	0.08	0.01	0.01
United States	0.04	0.33	0.01	0.01
Venezuela	0.07	0.75	0.06	0.67

Table 24 **Islamic Institutions in the GCC**

	Bahrain	Kuwait	Oman	Qatar	Saudi Arabia	UAE
Ban on Interest (in theory)	no	no	no	no	yes	no
Alms Tax	no	no	no	no	yes	no
Very Low Female Labor Force Participation/ High Birthrates	no	no	yes	partly	yes	partly
Public Morals	no	no	no	no	yes	no
Islamic Law's Influence on the Legal System	only PSL*	only PSL	only PSL	only PSL	strong	partly/Abu Dhabi
High Terror Threat	no	no	no	no	yes	no
Score (Fuzzy Value)	0 (0)	0 (0)	0.2 (0.2)	0.1 (0.1)	1.2 (1)	0.2 (0.2)

*PSL = personal status law

For the enlarged sample this list will be extended, but only Iran is comparable to Saudi Arabia in this respect. I will contrast the condition ISLAM with a standard variable on Islam: PERCENTMUSLIM is merely based on the percentage of Muslims in the population as of 1970.[20] PERCENTMUSLIM will also be tested as a variable in the regression. Its values can be taken from table 25.

Neither condition will be calibrated. The data is already convenient for fuzzy analysis and there is no theoretical assumption which would point to another calibration.

As for features of Arab societies, a common regression would probably use a dummy variable, assigning Arab countries a "1" and non-Arab countries a "0" (see table 25). I will test this practice with the variable ARAB in the regression and also with an equally constructed condition ARAB in the enlarged fsQCA sample. The basis was the membership in the League of Arab States. As for the small GCC sample, there obviously is a lack of variance for such a condition.

Democracy

As mentioned above, democracy is a frequently used although contentious variable in many growth regressions. When I conducted this study, the Freedom House Index was the only

Table 25 ISLAM, PERCENTMUSLIM, and ARAB

Country	ISLAM	PERCENTMUSLIM/ PERCENTMUSLIM	ARAB/ARAB
Algeria	0.2	0.97	1
Bahrain	0	0.95	1
Botswana	0	0	0
Brazil	0	0	1
Brunei	0.2	0	0
Canada	0	0	0
Ecuador	0	0	0
Egypt	0.1	0.81	1
Gabon	0	0.01	0
Indonesia	0.2	0.43	0
Iran	0.8	0.98	0
Iraq	0.2	0.95	1
Kuwait	0	0.4	1
Malaysia	0.2	0.5	0
Mexico	0	0	0
Netherlands	0	0	0
Nigeria	0.2	0.44	0
Norway	0	0	0
Oman	0.2	0.99	1
Qatar	0.1	0.95	1
Saudi Arabia	1	0.99	1
United Arab Emirates	0.2	0.92	1
United Kingdom	0	0.01	0
United States	0	0	0
Venezuela	0	0	0

index covering all sample countries in the entire time period of the analysis.[21] Although Freedom House is frequently criticized as biased, most researchers revert to it because of a lack of viable alternatives. One alternative is the Polity IV index (Polity IV Project/Center for Systemic Peace 2013), but this index does not cover as many countries and years as the Freedom House Index. (Hadenius/Teorell 2004) Freedom House provides assessments of a country's civil liberties (from 1 to 7) and political rights (from 1 to 7). In most cases, both values are highly correlated. Because economic rights (part of civil liberties) are considered sometimes as more important than political rights an attempt was made to create two different conditions, but in no case did the difference between both values play decisive roles. Hence, I opted for one single variable/condition "democracy". Raw data and the fuzzy values can be taken from table 26.
For the first condition (DEMOCRACY), the turning point is an average Freedom House value of 4. Full membership in the set of autocratic countries (fuzzy value 0) is assigned to countries with a score of 6 or higher. Full membership in the set of democratic countries (1) is reached when the Freedom House value is lower than 2. Fuzzy values from 0.5 to 0.9 indicate that a country has meaningful democratic structures, while fuzzy values between 0.1 and 0.49 indicate autocratic regimes.

For rentier states, the theoretical background given in chapter three made several specific assumptions that can lead to a different calibration for the first fsQCA analysis, covering only GCC states. A special calibration is nevertheless necessary because of the fact that all GCC countries are autocratic. Therefore, I propose a new condition on democracy, which results from a different calibration, based on a different theoretical reasoning.

From a theoretical point of view, I expect that, rather than allowing for a meaningful and constructive participation of the people, democratic organs are used as more sophisticated instruments for rent-seeking. When these organs have veto powers and have meaningful rights in the decision making process, they can impede necessary reforms and hamper growth. To test this hypothesis, I will introduce a condition called "PARLIAMENT", which is based on the Freedom House ratings for political rights only. The turning point for the calibration will be between consultative councils without veto power or without the possibility to make use of their theoretical rights, and parliaments with meaningful legislative rights. In the GCC sample, the turning point would be between Bahrain and Kuwait: 5. In the Freedom House ranking, this is the turning point between "partly free" (from 3 to 5) and "not free" (higher than 5). "PARLIAMENT" thus divides the GCC states into one partly free state and "unfree" states. Using insight from the case studies, these categories could also be identified as meaningful parliament (lower than 5), and consultative council without meaningful legislative rights. Please note that the data covers the period from 1973 to 2006. During most of this time, the Kuwaiti parliament was suspended. Looking at the average data from 1991 to 2006, for instance, the cleavage between Kuwait and the other GCC states would be even deeper as the data in table 27 suggests.

Table 26 Average Freedom House Scores

Country	Freedom House Civil Liberties and Political Rights, averaged (1973-2006)	Fuzzy Values "DEMOCRACY"
Algeria	5.75	0.07
Bahrain	5.33	0.12
Botswana	2.2	0.9
Brazil	3.06	0.76
Brunei	5.71	0.07
Canada	1	1
Ecuador	3.08	0.75
Egypt	5.29	0.13
Gabon	5.19	0.14
Indonesia	4.62	0.28
Iran	5.82	0.06
Iraq	6.72	0
Kuwait	4.63	0.28
Malaysia	4.07	0.47
Mexico	3.27	0.71
Netherlands	1	1
Nigeria	4.65	0.27
Norway	1	1
Oman	5.89	0.06
Qatar	5.63	0.08
Saudi Arabia	6.56	0
United Arab Emirates	5.42	0.11
United Kingdom	1.16	1
United States	1	1
Venezuela	2.59	0.84

Source: Freedom House.

Table 27 **Freedom House and PARLIAMENT**

	Freedom House: Political Rights (1973-2006)	Freedom House: Political Rights (1991-2006)	Fuzzy Sets PARLIAMENT
Bahrain	5.77	6.06	0.17
Kuwait	4.91	4.75	0.59
Oman	6.12	6	0.18
Qatar	5.94	6.56	0.09
Saudi Arabia	6.52	7	0.05
UAE	5.7	6.06	0.17

Source: Freedom House.

Institutional Quality

Douglas North's definition of institutions is very broad. As a consequence, NIE deals with many different kinds of institutions. Morality and religion could also be regarded as institutions because they can reduce transaction costs. (Brinitzer 2003) One aspect of institutional quality is already covered by the democracy variable/condition and this variable strongly overlaps with one of the World Bank's six governance indicators: "Voice and Accountability". The other five indicators [22] are defined in the methodological paper by Kaufmann, Kray, and Mastruzzi as follows:

2. Political Stability and Absence of Violence/Terrorism (PV) – capturing perceptions of the likelihood that the government will be destabilized or overthrown by unconstitutional or violent means, including politically motivated violence and terrorism.

(b) The capacity of the government to effectively formulate and implement sound policies:

3. Government Effectiveness (GE) – capturing perceptions of the quality of public services, the quality of the civil service and the degree of its independence from political pressures, the quality of policy formulation and implementation, and the credibility of the government's commitment to such policies.

4. Regulatory Quality (RQ) – capturing perceptions of the ability of the government to formulate and implement sound policies and regulations that permit and promote private sector development.

(c) *The respect of citizens and the state for the institutions that govern economic and social interactions among them:*

> **5. Rule of Law (RL)** – capturing perceptions of the extent to which agents have confidence in and abide by the rules of society, and in particular the quality of contract enforcement, property rights, the police, and the courts, as well as the likelihood of crime and violence.
>
> **6. Control of Corruption (CC)** – capturing perceptions of the extent to which public power is exercised for private gain, including both petty and grand forms of corruption, as well as "capture" of the state by elites and private interests. (Kaufmann/Kraay/Mastruzzi 2010: 4)

Political stability will form a single variable/condition called POLSTAB/POLSTAB. For the years before the governance ratings (before 1996) I created my own data using the criteria of the political stability ratings and qualitative information from the Polity IV Index. This, however, was only done for the 25 sample countries.

To reduce the number of variables/conditions, government effectiveness will be merged with regulatory quality. The resulting variable/condition will be called GOVREGULAR/GOVREGULAR. Both categories seem to be two sides of the same coin. Similarly, rule of law and control of corruption are related categories and will be merged into RULEOFLAW/RULEOFLAW.

Furthermore, there will be a fourth "meta" variable/condition resulting from the average of government effectiveness, regulatory quality, rule of law, control of corruption, and political stability: INSTITUTIONS. This variable is, as will be shown, necessary to avoid correlation of the regression variables due to these often co-variant institutional variables.

It should be noted that these variables/conditions only cover the period from 1996 to 2007. Raw data can be taken from table 28.

These figures are already aggregate indicators and based on an array of other indices. "All the individual variables have been rescaled to run from zero to one, with higher values indicating better outcomes." (Kaufmann/Kraay/Mastruzzi 2010: 8) The individual variables were then combined using a method called the unobserved components model. The results are indicators ranging roughly from -2.5 to 2.5 for given standard deviations, indicating that small differences between two countries should not be overrated. For the GCC states, standard deviations were all very similar.

This complex method and the fact that the six above-mentioned categories are each based on several indicators and data sources make it opportune to stick to these scales for fuzzy set calibration (see table 29). The "natural" turning point, which will receive the fuzzy value 0.5, is 0. For the fuzzy values 0 and 1, the variance in the sample was decisive. To reach 1 in political stability, 0.9 were necessary, while -0.9 led to a fuzzy value of 0. Good values for governance and regulatory quality (GOVREGULAR) were apparently harder to achieve. Here, 0.7 were necessary for 1 and -0.7 for 0. For rule of law and control of corruption (RULEOFLAW), 0 and 1 were attained at-0.8 and 0.8 respectively.

Table 28 Average World Bank Governance Ratings 1996-2007

Country	Political Stability, averaged from 1973-2007	Government Effectiveness	Regulatory Quality	Rule of Law	Control of Corruption
Algeria	-1.71	-0.6	-0.72	-0.85	-1.65
Bahrain	-0.84	0.54	0.8	0.63	0.59
Botswana	0.96	0.65	0.66	0.62	0.82
Brazil	-0.46	-0.07	0.17	-0.35	-0.1
Brunei	1.2	0.74	0.93	0.47	0.32
Canada	0.69	2	1.48	1.78	2.03
Ecuador	-0.92	-0.87	-0.59	-0.76	-0.83
Egypt	-0.85	-0.35	-0.34	-0.04	-0.38
Gabon	0.01	-0.62	-0.22	-0.55	-0.75
Indonesia	-0.51	-0.46	-0.37	-0.79	-0.9
Iran	-1.12	-0.61	-1.45	-0.65	-0.52
Iraq	-1.78	-1.77	-1.88	-1.7	-1.42
Kuwait	0.01	0.25	0.28	0.73	0.85
Malaysia	0.43	0.89	0.54	0.51	0.34
Mexico	-0.63	0.16	0.42	-0.47	-0.35
Netherlands	1.15	2.03	1.76	1.76	2.13
Nigeria	-1.65	-1	-1.01	-1.37	-1.21
Norway	1.3	2.06	1.33	3.07	2.14
Oman	0.53	0.54	0.62	0.78	0.68
Qatar	0.67	0.52	0.31	0.67	0.72
Saudi Arabia	-0.06	-0.26	-0.11	0.27	0.15
United Arab Emirates	0.71	0.74	0.71	0.82	0.9
United Kingdom	0.62	1.89	1.73	1.74	2.04
United States	0.65	1.77	1.49	1.6	1.66
Venezuela	-0.69	-0.73	-0.84	-1.09	-0.94

Source: World Bank.

Summary

The methodological approach can be summarized as follows: This study is based on a methodological triangulation consisting firstly of case studies of the GCC economies in chapter two. Secondly, the GCC states will be analyzed "within a sample of up to 181 countries, which will be the object of several regressions." Case studies and regressions are the small-n and and large-n cornerstones for the comparison of methodologies for the analysis of growth performances. Their results and their methodological implications will be compared to a third analytical method, based on fsQCA. Two fsQCA analyses will be made. In a first step, the GCC states will form a sample of 6 cases, while as a second step, the sample will be enlarged to 25 (26) cases by including further oil-rich countries.

FsQCA is a software-based method, enabling the researcher to find necessary and sufficient conditions, or in most cases, combinations of conditions that lead to a certain outcome. It does not work with raw data but with the membership scores in the fuzzy sets that have to be assigned to each country, and for each condition by the researcher, based on data, estimates and theoretical reasoning. A fuzzy value of 1.0 means that a country is fully a member in a certain set (for example the set of democratic countries), while lower fuzzy values indicate that it is only partly a member of this group.

The time frame of the analysis is from 1973 to 2006. The 25 (26) sample countries have been chosen from countries with significant oil resources but a special focus lies on the GCC countries.

Table 29 **Merged Variables/Conditions and Fuzzy Sets – Institutional Indicators**

Country	Political Stability, averaged from 1973-2007	Fuzzy Value	GOVREGULAR	Fuzzy Value	RULEOFLAW	Fuzzy Value
Algeria	-1.71	0	-0.73	0.04	-1.36	0
Bahrain	-0.84	0.06	0.66	0.94	0.53	0.88
Botswana	0.96	1	0.63	0.94	0.67	0.93
Brazil	-0.46	0.18	0.05	0.55	-0.2	0.32
Brunei	1.2	1	0.83	0.97	0.43	0.83
Canada	0.69	0.91	1.68	1	1.93	1
Ecuador	-0.92	0.04	-0.6	0.07	-0.76	0.05
Egypt	-0.85	0.06	-0.28	0.23	-0.16	0.35
Gabon	0.01	0.55	-0.37	0.17	-0.71	0.07
Indonesia	-0.51	0.15	-0.32	0.2	-0.8	0.05
Iran	-1.12	0.02	-1.07	0	-0.63	0.09
Iraq	-1.78	0	-1.92	0	-1.54	0
Kuwait	0.01	0.55	0.21	0.71	0.8	0.95
Malaysia	0.43	0.81	0.7	0.95	0.47	0.85
Mexico	-0.63	0.11	0.3	0.78	-0.44	0.16
Netherlands	1.15	1	1.88	1	1.96	1
Nigeria	-1.65	0	-1.05	0	-1.28	0
Norway	1.3	1	1.69	1	2.44	1
Oman	0.53	0.85	0.61	0.93	0.71	0.93
Qatar	0.67	0.9	0.42	0.86	0.56	0.89
Saudi Arabia	-0.06	0.45	-0.22	0.28	0.21	0.69
United Arab Emirates	0.71	0.91	0.68	0.95	0.79	0.95
United Kingdom	0.62	0.89	1.82	1	1.93	1
United States	0.65	0.9	1.64	1	1.67	1
Venezuela	-0.69	0.09	-0.59	0.07	-0.93	0.03

Table 30 summarizes the variables and conditions which will be used, while some of them will only be used for either the regression or the fsQCA-analysis.

Table 30 **Theory, Variables/Conditions and Outcome**

	Theory	Independent Variables/ Conditions	Dependent Variable/ Outcome
First Bloc	***Growth Theory***	Initial Non-Oil GDP (LOGNDINITIAL)	*Non-Oil Growth* (NDGROWTH)
		Investment (LOGINVEST, INVESTMENT, INVESTCAPITA)	
		Human Capital (AVSCHOOLING, HUMANCAP)	
		Population Growth (FERTILITY, FERTILITY)	
Second Bloc	***Dutch Disease***	Amount of the Income from Resources (*per capita* and per year) (RENT, RENT)	
	Rentier State Theory	Rents as a Percentage of Government Income (RENTIER, RENTIERI, RENTIERII)	
	Incentives (Empirics)	Natural Resource Reserves *per capita* (national) (RESERVES, RESERVES)	
	Islam	Islamic Institutions (ISLAM) Muslims as a Percentage of the Population (PERCENTMUSLIM, PERCENTMUSLIM)	
	Arab Society	Dummy Variable/Dummy-Style Condition (ARAB, ARAB)	
	Democracy	Democracy (DEMOCRACY, DEMOCRACY) Meaningful Parliament/Elections (PARLIAMENT)	
	NIE	Political Stability (POLSTAB, POLSTAB)	
		Regulatory Quality and Governance (GOVREGULAR), Rule of Law and Absence of Corruption (RULEOFLAW) or an Average of POLSTAB, GOVREGULAR, and RULEOFLAW, INSTITUTIONS RULEOFLAW → INSTITUTIONS	

The choice and also the operationalization of the variables/conditions meant a lot of compromises. Several conditions for growth cannot be operationalized appropriately even if theories highlight their importance. Elite constellations, for instance, are such a case. Some conditions cannot be used for the small fsQCA sample because there is no meaningful variance. Moreover, the number of conditions has to be restricted due to the small sample sizes. All these problems make clear that these analyses cannot be made mechanically and must be prepared and interpreted with care.

Ultimately every operationalization means a loss of information. The attribute "qualitative" in QCA therefore must not mislead in that QCA inevitably also suffers from generalizations.

5

Empirical Analysis and Results

QCA is supposed to bridge the gap between qualitative and quantitative methods. The purpose offers certain advantages in comparison to regression analysis and is reflected in the literature on fsQCA. (Ragin 2006) This is why I will start this analytical chapter with a cross-country growth regression, using the most common uncalibrated variables in the largest possible sample. The results will allow for a comparison of results, and of methodological advantages or disadvantages of QCA in particular.

I will then proceed to the fsQCA analysis of the six cases of the states of the Gulf Cooperation Council and of the larger sample of countries with important resource rents. The methodological chapter has demonstrated the complexity of variable/condition choices and their relation to the sample countries and sample size. Therefore, table 31 gives an overview of the variables and conditions used for the different analyses to be conducted in this chapter. To better differentiate between variables and conditions, variable names will appear in capital letters, while condition names are in small caps.

5.1 Regressions

In this first model, I deliberately follow the common practice of cross-country regressions with the exception of a few variables, which I operationalized differently. This means that insight gained from case studies of certain countries will not be considered because regressions tend to discuss theories or previous models but rarely cases. There is also a necessity to find one operationalization for all cases – sometimes despite knowledge of the inappropriateness of this operationalization for certain cases.

The model is a multivariate OLS-regression based on data which has been averaged from 1973 to 2006 (or slightly earlier for independent variables). The dependent variable is non-oil growth from 1973 to 2006 (NDGROWTH). A model based on panel-data has been rejected due to massive data problems affecting the dependent variable in

Table 31 Overview of Variables and Conditions in the Different Models

	Regression Analysis	FsQCA Analysis (small sample)	FsQCA Analysis (large sample)
Sample Sizes	111-181	6 countries (7 cases)	25 countries (26 cases)
Initial GDP	LOGNDINITIAL (GDP minus resource rents, natural log)	dropped (included in the calibration of the outcome)	dropped because included in the calibration of the outcome
Investments	LOGINVEST (investment share of GDP, natural log)	INVESTCAPITA (investment *per capita* relative to rent *per capita*)	INVESTMENTS (investment share of GDP)
Human Capital	AVSCHOOLING (average school enrollment, years)	dropped (unreliable data)	HUMANCAP
Women	WOMENLAB, (percentage of women in the labor force)	included in ISLAM	WOMENLAB (percentage of women in the labor force)
Fertility	FERTILITY	included in ISLAM	FERTILITY
Natural Resources			
Resource Abundance	RENT	RENT	RENT
Resource Abundance	RESERVES	RESERVES	RESERVES
Resource Dependence	SXP	SXP	SXP
Resource Dependence	RENTIER	RENTIERII	RENTIERI
Islam	PERCENTMUSLIM	ISLAM	ISLAM
Arab Society	ARAB (dummy)	ARAB (crisp set)	no variance
Democracy	DEMOCRACY (Freedom House reversed)	DEMOCRACY (Freedom House reversed)	PARLIAMENT
Institutions	INSTITUTIONS (average of POLSTAB, GOVREGULAR and RULEOFLAW)	POLSTAB; GOVREGULAR; RULEOFLAW;	POLSTAB; GOVREGULAR; RULEOFLAW;

particular. While averaged data is likely to grasp the overall tendencies of the variables, the single data points are in many cases highly irrational and improbable. This is especially true for "non-oil" data. A time series data analysis would over-interpret changes from one year to the next.

The independent variables are initial non-oil GDP (NDINITIAL), investment as a share of GDP (INVEST), fertility rate (FERTILITY) and a measure of human capital (AVSCHOOLING). INVEST and NDINITIA) enter the regression as natural log (LOGNDINITIAL, LOGINVEST) because a graphical illustration showed a logarithmic shape as in many growth regressions. FERTILITY and female labor force participation rates (WOMENLAB) have a similar theoretical background and a choice should be made between the two. FERTILITY covers more countries and data is probably more reliable: Female labor force participation can easily be distorted by labor migration, and data sources are often not clear about this issue. Furthermore, fertility rates could have more explanatory power given that the dependent variable is *per capita* growth. Testing both, fertility rates were significant, while WOMENLAB was not.

Islam will enter the regression as a percentage of the population: Despite reservations about this kind of operationalization, I deliberately followed this common practice. To avoid confusion with the condition on Islamic economic institutions (ISLAM), the variable is called PERCENTMUSLIM.

As for democracy (DEMOCRACY), I reversed the Freedom House scores and the highest scores now go to the most democratic countries.

Three variables represent institutions. Political stability (POLSTAB) was taken from the World Bank rating. The two World Bank ratings for regulatory quality and government effectiveness were aggregated to form a new "governance" (GOVREGULAR) variable, while rule of law and absence of corruption have been subsumed under RULEOFLAW. These institutional variables were all significant in the model, but they turned out to be highly correlated as table 32 shows. This is why the average of all three variables was calculated for a new aggregated variable: INSTITUTIONS.

***Table 32* STATA Output – Correlation of Institutional Variables**

	POLSTAB	RULEOFLAW	GOVREGULAR
POLSTAB	1.0000		
RULEOFLAW	0.8095*	1.0000	
GOVREGULAR	0.7464*	0.9509*	1.0000

* Significant correlation on a 0.05 level.

The first basic model consists therefore of NDGROWTH as a dependent, and LOGNDINITIAL, LOGINVEST, FERTILITY, INSTITUTIONS, PERCENTMUSLIM, ARAB and AVSCHOOLING as independent variables. (see equation 9)

Equation 9:

$$NDGROWTH_i = \beta 0 + \beta_i \, LOGNDINITIAL + LOGINVEST + \beta_i \, INSTITUTIONS \; \beta_i \, FERTILITY + \beta_i \, PERCENTMUSLIM + \beta_i \, ARAB + \beta_i \, AVSCHOOLING + u +$$

The results of the regression can be taken from table 33.

Table 33 STATA Output – Result for the Basic Model

<table>
<tr><td>Source</td><td>|</td><td>SS</td><td>df MS</td><td>Number of obs</td><td>= 156</td><td></td></tr>
<tr><td></td><td></td><td>F(7, 148)</td><td>= 20.12</td><td></td><td></td><td></td></tr>
<tr><td>Model</td><td>|</td><td>351.202784</td><td>7 50.1718263</td><td>Prob > F</td><td>= 0.0000</td><td></td></tr>
<tr><td>Residual</td><td>|</td><td>368.975464</td><td>148 2.49307746</td><td>R-squared</td><td>= 0.4877</td><td></td></tr>
<tr><td></td><td></td><td>Adj R-squared</td><td>= 0.4634</td><td></td><td></td><td></td></tr>
<tr><td>Total</td><td>|</td><td>720.178248</td><td>155 4.64631128</td><td>Root MSE</td><td>= 1.5789</td><td></td></tr>
<tr><td>ndgrowth735</td><td>|</td><td>Coef.</td><td>Std. Err. t</td><td>P>|t|</td><td>[95% Conf.</td><td>Interval]</td></tr>
<tr><td>loginvest</td><td>|</td><td>.8291538</td><td>.2778341 2.98</td><td>0.003</td><td>.2801197</td><td>1.378188</td></tr>
<tr><td>logndinitial</td><td>|</td><td>-1.875361</td><td>.2399317 -7.82</td><td>0.000</td><td>-2.349495</td><td>-1.401226</td></tr>
<tr><td>institutions</td><td>|</td><td>1.00571</td><td>.2166807 4.64</td><td>0.000</td><td>.5775225</td><td>1.433898</td></tr>
<tr><td>avschooling</td><td>|</td><td>-.0751232</td><td>.0865482 -0.87</td><td>0.387</td><td>-.2461531</td><td>.0959066</td></tr>
<tr><td>percentmus~m</td><td>|</td><td>.0979348</td><td>.214463 0.46</td><td>0.649</td><td>-.3258704</td><td>.5217401</td></tr>
<tr><td>fertility</td><td>|</td><td>-.8867828</td><td>.1390793 -6.38</td><td>0.000</td><td>-1.16162</td><td>-.6119451</td></tr>
<tr><td>arab</td><td>|</td><td>.5226607</td><td>.476465 1.10</td><td>0.274</td><td>-.4188925</td><td>1.464214</td></tr>
<tr><td>_cons</td><td>|</td><td>18.46381</td><td>2.250852 8.20</td><td>0.000</td><td>14.01586</td><td>22.91177</td></tr>
</table>

The results are four fairly robust variables: LOGNDINITIAL, LOGINVEST, FERTILITY, and INSTITUTIONS all display the expected significant correlations with

NDGROWTH. A lower initial GDP correlates with higher growth rates, as convergence theory suggests. High fertility rates are detrimental to (*per capita!*) growth, while investments and good institutions are correlated positively with growth.

AVSCHOOLING, ARAB and PERCENTMUSLIM are not significant. A pairwise correlation test shows that the average years of schooling (AVSCHOOLING) correlates with FERTILITY. Indeed, AVSCHOOLING turns significant (t 2.63, p 0.009, Coeff. 0.218) if FERTILTIY is not added to the model, but the robustness of FERTILITY is higher. I decided to drop AVSCHOOLING and not FERTILITY, partly due to the above mentioned operationalization problems of human capital. For the sake of parsimonity, I also dropped PERCENTMUSLIM and ARAB – these variables had been added to follow a widespread practice. The theoretical background of the underlying operationalization, however, could not be agreed upon and their insignificance only confirms what could be assumed from the approach to measure the impact of cultural or religious phenomena on economic growth.

To this basic model of now four control variables, I added the more disputed variables: DEMOCRACY, and separately, the "resource variables": The annual average resource rent *per capita* (RENT), the amount of reserves *per capita* (RESERVES), the share of primary product exports in GNP (SXP), and the share of rents in government income (RENTIER). The results (models 1-5) can be taken from table 34.

To begin with, DEMOCRACY is not significant. The heterogeneous findings of other regressions and an ambiguous theoretical background gave reason to expect this result.

Contrary to earlier findings, the "resource variables" also fail to be significant. This is even true for the resource dependence variables which often turned out to be highly significant in previous regression models by other authors. RENTIER is significant on a 0.05 confidence level, but *positively* correlated with growth. In an earlier regression, based on PENN World Table 6.2 data (instead of 6.3 data), RENT correlated positively with growth but these findings cannot be considered robust. (Kropf 2010) Hence, also the positive correlation of resource abundance with growth, as found by Brunnschweiler (2008), could not be reproduced. This is not astonishing given that the dependent variable in my regression is non-oil growth. In Brunnschweiler's regression, the dependent variable is based on normal growth rates.

It is also important to note that sample sizes vary according to data availability. The share of natural resource exports in GNP (SXP) especially fails to cover the entire sample: The sample size of model 4 is thus considerably smaller than in the other models and the comparability is restricted.

I tested every model for heteroscedasticity. Applying the Breusch-Pagan/Cook-Weisberg test, heteroscedasticity could always be rejected.

Table 34 **Overview – Results of the five Models (Coefficients, Standard Errors in Parentheses)**

	1. Model	2. Model	3. Model	4. Model	5. Model
logndinitial	-1.730383**	-1.874964**	-1.748131**	-2.110687**	-1.918017**
	(.2045086)	(.2244981)	(.2280973)	(.2532507)	(.2086846)
loginvest	1.03027**	.965557**	1.01512**	.910015**	.8583573**
	(.2691525)	(.2714234)	(.2830395)	(.3209764)	(.2642133)
fertility	-.7506968**	-.7971693**	-.7534367**	-.936796**	-.7641142**
	(.1156108)	(.1190908)	(.1169524)	(.1667579)	(.1290468)
institutions	1.122198**	1.024945**	1.122664**	1.269429**	1.099868**
	(.2572649)	(.2640145)	(.2579929)	(.2511532)	(.2380605)
democracy	-.1836837	-.101213	-.175385	-.2321096	-.0940219
	(.1129145)	(.1246856)	(.1224773)	(.1456123)	(.1142466)
rent		.1712255			
		(.1116988)			
reserves			.0728007		
			(.4096007)		
sxp				.1936969	
				(.9650506)	
rentier					1.487215*
					(.6826678)
Obs.	181	181	181	111	139
Adj. R^2	0.4263	0.4307	0.4231	0.5743	0.5189

** significant on a 0.01 confidence level
* significant on a 0.05 confidence level

Theory also suggests that natural resources and democracy (or rather autocracy) are correlated. Rentier state theory is based on the finding that rentier states are mostly autocratic. Therefore, interaction terms of democracy on the one hand and different resource variables on the other hand were introduced, while keeping the respective "single" variables in the model. None of these interaction terms was significant as table 35 shows. In the case of RENTIER, the introduction of the RENTIER*DEMOCRACY interaction term resulted in the loss of significance of the RENTIER variable.

These regressions, while reflecting widespread practice, deviate from former analyses in that:

- No outliers were excluded;
- The sample size of 181 (at maximum), for instance, is more than twice as large as the sample size of the Sachs and Warner regressions;

- The time frame was extended to 2005/2006;
- Non-oil growth rates have been used.

Table 35 **Interaction Terms – DEMOCRACY and Resource Variables**

	Coef. (Std. Err)	**T (p-value)**	**Behavior of Democracy and Resource Variable**
INTERACTRENT	.0014792 (.00414119)	0.36 (0.721)	RENT/DEMOCRACY not significant
INTERACTRESERVES	.0130259 (.0153861)	0.85(0.398)	RESERVES/ DEMOCRACY not significant
INTERACTSXP	0.0758106 (.0425068)	1.78 (0.078)	SXP/DEMOCRACY not significant
INTERACTRENTIER	-.2201262 (.4988184)	-0.44 (0.660)	RENTIER/ DEMOCRACY not significant

The fact that resource variables turned out to be significant in other models may be attributed to those differences. If anything, this model shows how easily results from previous regressions can be questioned.

Nevertheless, RENT, RESERVES, SXP, and RENTIER must not be lumped together as entirely meaningless, even if they all fail to be significant in this model. There are differences between these variables and even the correlation among them is not as high as could be assumed as table 36 shows:

Table 36 **STATA Output – Correlation Test**

	RENT	RESERVES	SXP	RENTIER
RENT	1.0000			
RESERVES	0.7183	1.0000		
SXP	0.5792	0.5324	1.0000	
RENTIER	0.7106	0.7398	0.6220	1.0000

As I have demonstrated elsewhere, the use of a different resource variable may make a difference: The replacement of SXP by RENT in a model by Sachs/Warner models, for example, has resulted in the loss of significance of the resource variable. (Kropf 2010)

Furthermore, testing the resource variables in a model explaining democracy, their significance is much less contentious. In a very simple model testing the dependent variable democracy and the log of non-oil initial GDP (LOGNDINITIAL) as independent variables, fertility rates (FERTILITY), political stability (POLSTAB), "Arabness" (ARAB), the percentage of Muslims in the entire population (PERCENTMUSLIM), and the resource variables, some of the insignificant variables from above are significant. (See table 37) In all four regressions, ARAB is significant and negatively correlated to DEMOCRACY, while PERCENTMUSLIM is not. And, as in many regressions, RENT (model 2a in table 37), RESERVES (model 3a in table 37), and RENTIER (model 4a in table 37) turned out to be significant and negatively correlated to democracy. SXP, interestingly, is not significant (model 5a in table 37). Nevertheless, this model is far from inclusive of a sufficient number of potential control variables.

***Table 37* DEMOCRACY as Dependent Variable (Coefficients, Standard Error in Parentheses)**

	Model 1a	Model 2a	Model 3a	Model 4a	Model 5a
LOGNDINITIAL	0.61 (.13)*** p= 0.000	0.82 (.13)*** p= 0.000	0.82 (.13)*** p= 0.000	0.55 (.14)*** p=0.000	0.53 (.14)*** p=0.000
FERTILITY	-0.17 (.08)** p= 0.023	-1 (.72) p= 0.19	-0.14 (.07)** p= 0.055	-0.24 (.08)*** p=0.005	-0.32 (.08)*** p=0.000
POLSTAB	0.69 (.12)*** p= 0.000	0.75 (.11)*** p= 0.000	0.64 (.11)*** p= 0.000	0.5 (.12)*** p=0.000	0.49 (.11)*** p=0.000
ARAB	-1.44 (.31)*** p= 0.000	-0.88 (.32)*** p= 0.007	-1.04 (.31)*** p=0.001	-1.29 (.33)*** p=0.000	-1.34 (.38)*** p=0.001
PERCENTMUSLIM	-0.2 (.16) p= 0.922	0.03 (.15) p= 0.843	0.04 (.15) p=0.778	0.09 (.14) p=0.528	-0.37 (.35) p=0.291
RENT		-0.34 (.69)*** p= 0.000			
RESERVES			-1.25 (0.23)*** p= 0.000		
RENTIER				-1.38 (0.48)*** p=0.005	
SXP					-0.74 (0.61) p=0.227
Adj. R-squared	0.61	0.66	0.66	0.67	0.81
Sample size	182	182	182	139	111

*** Significant in a 0.01 confidence interval

** Significant in a 0.05 confidence interval

Thus, while democracy and resources seem to explain very little in my first regressions (Model 1-5 in table 34), there is no reason to assume that these variables are meaningless. The effect, however, may be highly dependent on other factors or it may only concern specific sub-sets. This is also indicated by the measure of fit of the entire model, which is not particularly high. There is reason to believe that this is due to the sample size. Reducing, for example, my sample size to the cases covered by SXP, R^2 is considerably higher. The adjusted R^2 is highest in the fourth model which has the smallest sample size. Higher sample sizes come along with lower R^2. (See table 34 above)

What conclusion can be drawn from this study for the GCC countries? I would argue that none can be. As in earlier regressions, the residuals of the GCC states are rather high and two of them would probably be excluded as outliers, while only Oman and Qatar have low residuals. (See table 38) This is remarkable: While the GCC states (and other Gulf States) are resource abundant and dependent states par excellence, empirical investigation of the resource curse on an international cross-country level fails to explain their growth performance. It is even possible that these six cases are not at all represented by the overall results of the regression, which is always possible for single cases or small sub samples in regressions, no matter how the model is adapted. Hence, there is one more reason to look at these states from another methodological angle.

Table 38 **Residuals of the GCC States**

	1. Model (DEMOCRACY)	2. Model (RENT)	3.Model (RESERVES)	4. Model (SXP)	5. Model (RENTIER)
Bahrain	6.7479038	6.4925642	6.7694325	6.8289485	6.4288373
Kuwait	-2.6315126	-3.5542071	-2.7072055	-2.2306452	-3.2668459
Oman	.65386915	.71921682	.62016088	.84649312	.00421948
Qatar	.35793832	-.85385525	.29277739	n.a.	-.10713919
Saudi Arabia	-2.5253761	-2.6282835	-2.5462151	-1.700546	-2.8117473
UAE	3.1474638	1.9804332	3.0646155	3.2093105	2.5656741

Note: Sachs and Warner did not provide a SXP value for Qatar.

5.2 FsQCA-Analysis of the GCC States

The regression analysis has shown how sets of countries can "disappear" in larger samples. Now, the sample size is considerably smaller. While there is no limit to the number of conditions that can be tested as necessary conditions, testing sufficient conditions should be restricted to fewer conditions. Here, priority shall be given to the contentious vairables on natural resources and DEMOCRACY.

An fsQCA analysis is a work in progress. Every case and every condition can be investigated closely. Conditions can be adapted to better reflect the background of the GCC economies. It is an important step in fsQCA to look at the fuzzy data more closely before and during the analysis. Being aware of the data situation can avoid misinterpretations of results. Frequently, problems result from covariance or limited variance. In some cases, even a recalibration may be justified.

I will shortly repeat the restrictions for the smaller sample, which have already been mentioned in the methodological chapter. However, several problems will only appear during the analysis, and how this happens can not be anticipated.

Due to the rentier multiplier effect, two factors are problematic: NDINITIAL and INVEST. Initial *per capita* GDP, even if resource rents are detracted, does not reflect the stage of development in the sense of convergence theory. The initial GDP variable clearly reflects resource rents and there is no simple solution to correct for this while keeping a meaningful variance at the same time: NDINITIAL will be dropped but had already been used for the calibration of the outcome condition. A similar problem occurs with respect to the investment condition. As explained in the methodology chapter, I will resort to the condition INVESTCAPITA. Furthermore, PARLIAMENT will replace the democracy conditions.

Also the conditions on fertility rates or women's participation in the labor force (see table 39) are almost covariant if the fuzzy values from the methodological chapter are used. Therefore they will be dropped while the "gender factor" is partly included in the condition ISLAM, which does not necessarily assume a connection of the two.

***Table 39* Fuzzy Values for Fertility Rates and Female Labor Force Participation**

	Births per Woman	Fuzzy Value	Female Labor Force Participation	Fuzzy Value
Bahrain	3.63	0.87	26.4	0.06
Kuwait	3.75	0.9	35.49	0.29
Oman	5.75	1	17.94	0.03
Qatar	4.33	0.98	29.73	0.11
Saudi Arabia	5.66	1	14.64	0
UAE	4.02	0.96	27.29	0.03

Furthermore, the fuzzy values for RENTIERI are all well above 0.5. The turning point was set at 40 percent, based on the definition of the allocation state by Luciani (Luciani 1987: 13). This turning point was meant to distinguish full rentier states from semi-rentiers. However, in the GCC, there are only full rentiers and the high fuzzy values are consistent with the theory. One may, however, ask if this calibration is *useful*. Looking at the raw data, there is a clear difference between Bahrain (59 percent of government revenue from rents) and the other GCC states (ranging from 0.8 to 0.88). This gap is significant and probably not a result of data errors: It also reflects qualitative findings. Thus, it makes sense to reconsider and change the calibration, creating a second "rentier" condition: RENTIERII. Contrary to RENTIERI that divides rentiers from semi-rentiers, RENTIERII divides rentier states into extremely oil-dependent and oil-dependent states. These differences are reflected in the new calibration in table 40.

Table 40 **RENTIERI and RENTIERII**

	Raw Data: RENTIER	RENTIERI	RENTIERII
Bahrain	0.59	0.62	0.25
Kuwait	0.88	0.98	0.94
Oman	0.81	0.96	0.84
Qatar	0.87	0.98	0.93
Saudi Arabia	0.8	0.95	0.82
UAE	0.84	0.97	0.89

The following table (41) summarizes the fuzzy sets for the GCC states, resulting from the above-mentioned restrictions.

Intercorrelation of variables does not pose a problem to fsQCA. Hence, the conditions GOVREGULAR and RULEOFLAW can be used separately and there is no need for the aggregate variable INSTITUTIONS as in the regression. It is noteworthy that, in this smaller sample, RULEOFLAW and GOVREGULAR are not covariant, which means that a low fuzzy value of GOVREGULAR can be found to be combined with a high value of RULEOFLAW. Also POLSTAB differs from RULEOFLAW and GOVREGULAR and not only due to the calibration. However, all GCC states have RULEOFLAW fuzzy values higher than 0.5 and the condition will have little explanatory power in this sample. Thus, the important institutional condition will be GOVREGULAR.

A closer look at the dataset also reveals that the condition ISLAM bears a similarity to the condition GOVREGULAR: The only country with high membership scores in the set of countries with important Islamic economic institutions (ISLAM) is Saudi Arabia. On the other hand, Saudi Arabia is the only country with fuzzy values of GOVREGULAR below 0.5. Given that ISLAM is based on institutions, this is not surprising. The correlation is not necessarily causation however.

In this small sample, there is also a noticeable covariance between RENT and RESERVES: Those countries with ample reserves also receive ample rents. While this is not necessarily true in the bigger sample, the different implications of two different conditions cannot be contrasted in this sample.

Table 41 Fuzzy Sets for the GCC Sample

	NDGROWTH	INVESTCAPITA	PARLIAMENT	POLSTAB	RULEOFLAW	GOVREGULAR
Bahrain	1	0.65	0.17	0.06	0.88	0.94
Kuwait	0	0.08	0.59	0.55	0.95	0.71
Oman	0.8	0.83	0.18	0.85	0.93	0.93
Qatar	0.1	0.23	0.09	0.9	0.89	0.86
Saudi Arabia	0	0.14	0.05	0.45	0.69	0.28
UAE	1	0.98	0.17	0.91	0.95	0.95

	ISLAM	RENT	RESERVES	SXP	RENTIERII
Bahrain	0	0.39	0.23	0.18	0.25
Kuwait	0	0.99	0.97	0.88	0.94
Oman	0.2	0.2	0.4	0.99	0.84
Qatar	0.1	1	1	0.95	0.93
Saudi Arabia	1	0.49	0.43	0.6	0.82
UAE	0.2	1	0.9	0.9	0.89

Test for Necessary Conditions

QCA analyses start with a test for necessary conditions. These conditions *have* to be present for a positive and negative outcome, no matter in which configuration of conditions. To this end, all conditions have to be tested as present and absent (the absence is indicated by a preceding swung dash: ~). For a necessary condition, consistency should be at least 0.9, while only higher consistencies justify to exclude a necessary condition from the following test for sufficient conditions.

The conditions were tested separately and the results can be taken from table 42.

Table 42 **Test for Necessary Conditions: GCC Sample (6 cases)**

Conditions tested	For NDGROWTH		For ~NDGROWTH	
	Consistency	Coverage	Consistency	Coverage
INVESTCAPITA	*0.87*	0.87	0.21	0.22
~INVESTCAPITA	0.22	0.21	*0.88*	0.88
POLSTAB	0.64	0.5	0.68	0.56
~POLSTAB	0.44	0.56	0.4	0.55
RULEOFLAW	**0.94**	0.52	*0.88*	0.52
~RULEOFLAW	0.12	0.48	0.17	0.76
GOVREGULAR	**0.96**	0.6	0.66	0.44
~GOVREGULAR	0.1	0.21	0.39	0.92
ISLAM	0.17	0.33	0.42	0.87
~ISLAM	**0.93**	0.6	0.68	0.47
PARLIAMENT	0.21	0.49	0.29	0.73
~PARLIAMENT	*0.88*	0.54	0.79	0.52
RENT	0.58	0.42	0.83	0.63
~RENT	0.49	0.73	0.23	0.37
RESERVES	0.56	0.41	0.81	0.64
~RESERVES	0.5	0.71	0.26	0.39
SXP	0.68	0.44	0.83	0.57
~SXP	0.34	0.65	0.19	0.39
RENTIERII	0.7	0.44	**0.92**	0.61
~RENTIERII	0.38	0.82	0.15	0.35

Four conditions turn out to be necessary: The highest consistency (0.96) is obtained when testing GOVREGULAR. Furthermore, RULEOFLAW would be a necessary condition if it had a meaningful variance, which it does not have. Interestingly, the absence of Islamic institutions (~ISLAM) is also necessary for a positive outcome. This finding does not come as a surprise: ISLAM is complementary to GOVREGULAR. Saudi Arabia, the only country with low GOVREGULAR, is also the only country with high ISLAM.

As the only necessary condition for a negative outcome (~NDGROWTH), RENTIERII reaches a consistency of 0.92. In brief, this means that those countries that failed to diversify were all high-level rentiers. However, among the good performers, there was one high level rentier too (Oman).

Other conditions came very close to necessary conditions: INVESTCAPITA obtained a consistency of 0.87, combined with a very high coverage. The absence of INVESTCAPITA (~INVESTCAPITA), in turn, comes even closer to a necessary condition for ~NDGROWTH (0.88), with an equally high coverage. Hence, the suspicion arises that the outcome and INVESTCAPITA may be covariant. This suspicion is confirmed by the calibrated data in table 43:

Table 43 **NDGROWTH and INVESTCAPITA**

	NDGROWTH	INVESTCAPITA
Bahrain	1	0.65
Kuwait	0	0.08
Oman	0.8	0.83
Qatar	0.1	0.23
Saudi Arabia	0	0.14
UAE	1	0.98

While most values are very close to each other, there is a considerable gap between INVESTCAPITA and NDGROWTH in Bahrain. Apparently there is no 1:1 translation of good investment decisions into growth. The Bahraini exception is also reflected in the xy-plot (figure 4). Most points are very close to the diagonal. If all points were on the diagonal (or very close to it, depending on consistency thresholds), the condition would be both necessary and sufficient for the outcome. Bahrain, however, is not close enough to the diagonal.

Even if the consistencies are only at 0.87 and 0.88 respectively, it is very likely that the condition INVESTCAPITA will make the analysis trivial as a condition that is both sufficient and necessary. As such, it could be the object of interpretation but excluded from further analysis. In this case, it remains to be seen how INVESTCAPITA behaves in the analysis of sufficient conditions.

Figure 4 FsQCA Plot INVESTCAPITA as a Necessary Condition

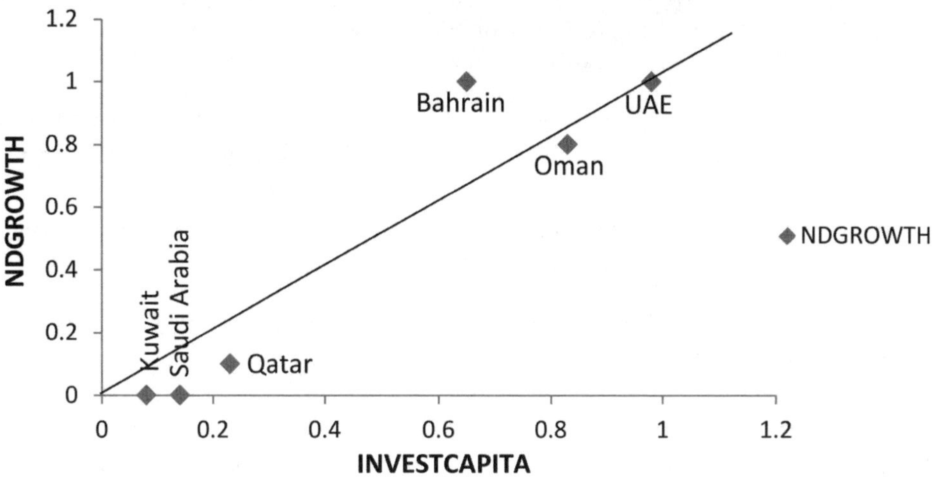

Another condition is almost necessary: The absence of a meaningful parliament (~PARLIAMENT) reaches a consistency of 0.88. No successful diversifier had a meaningful parliament but again this holds only true for a rough division of the sample along the 0.5 line. As for smaller variances of the fuzzy sets, relationships may vary, as figure 5 shows.

Figure 5 FsQCA Plot ~PARLIAMENT as a Necessary Condition

Summing up the necessary conditions, the result looks as follows:

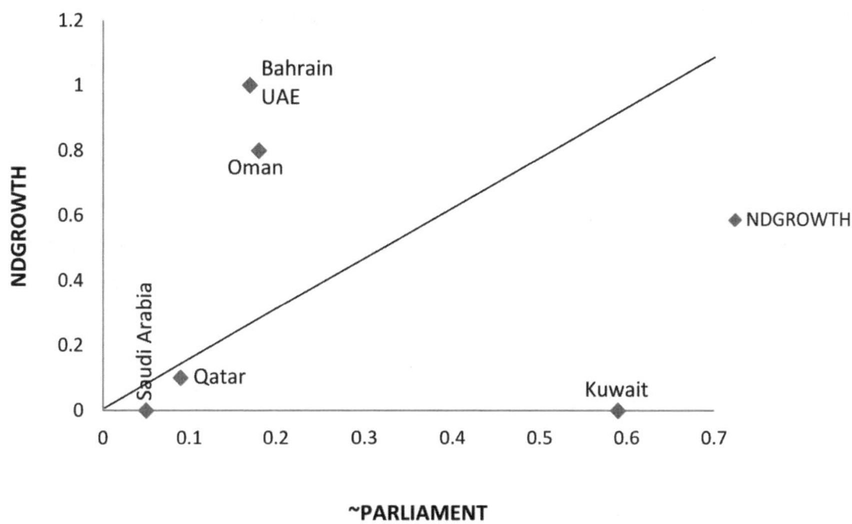

GOVREGULAR + ~ISLAM ← NDGROWTH

Good governance and regulatory quality ratings as well as the absence of Islamic economic institutions are necessary for a good diversification result.

RENTIERII ← ~NDGROWTH

Equally, being extremely resource dependent in terms of government income is apparently necessary for a failure to diversify. Such a result, of course, is highly dependent on the sample.

A test for GOVREGULAR*~ISLAM for NDGROWTH shows that also the following term is true:

GOVREGULAR * ~ISLAM ← NDGROWTH

As explained above, this is hardly astonishing as there is only one case that is not member in ~ISLAM*GOVREGULAR: Saudi Arabia.

Interpretation of Necessary Conditions

Among the results, the importance of regulatory quality and governance (GOVREGULAR) is the most striking. In all GCC states it is the government that plans economic development. In a rather short period of time, governments had to provide structures for the market, legal backgrounds, physical infrastructure, incentives for investment and investments themselves. If capacities for regulation and policy implementation are low, these efforts bear little fruit. Interestingly, such a low regulatory quality is found in Saudi Arabia. Saudi Arabia is the biggest of the GCC states, hence the challenge to administer and regulate such a large economy is bigger than in other GCC states compared to city-states. What is more, the Saudi Arabian government sometimes shied away from clear legal regulations that would have been convenient for investors but contradictory to the local interpretation of Islam. Furthermore, Kiren Aziz Chaudhry found that the Saudi Arabian government was faced with a powerful merchant opposition when reforms would have been necessary in the 1980s. (Chaudhry 1994) Similarly, Kuwait is known for its powerful merchant class and received a comparatively low rating of governance and regulatory quality, even if still rated positively by the fuzzy set calibration.

As for an extreme reliance on natural resource rents for government income (RENTIERII), it is difficult to draw such a net conclusion. From a theoretical background, it is unclear why RENTIERII out of all conditions should have turned out to be necessary for ~NDGROWTH. The small sample may account for this and the result may be somewhat of a coincidence. I will postpone the interpretation to the next chapter for reasons given below.

Also the fact that the absence of a meaningful parliament (~PARLIAMENT) came close to a necessary conditions calls for an explanation. Firstly, it has to be noted that there is very little variance in the sample: Among the GCC states only Kuwait has a meaningful parliament, even if Bahrain also has a parliamentary tradition. The Bahraini government, however, seems to be much more prone to resorting to violent suppression than the Kuwaiti government, and it overruled forces in parliament and religious figures whenever it deemed necessary. To put it bluntly: Bahrain's government dares to put political harmony at risk.

On the contrary, with its ample oil reserves, Kuwait can afford to perpetuate privileges as demanded by parliamentarians. Furthermore, the Iraqi invasion has paved the way for Kuwait's parliament to become stronger. Kuwait's strong merchant class has different channels to maintain its privileges and prevent reforms. There are thus several reasons why the Kuwaiti parliament has been given more power than the *majlis* in Bahrain. However, this power is not always used for the best of the economy. Experiences of Kuwaiti elections have shown that there is a tendency to vote for conservative Islamic candidates and traditional tribal leaders, while only few liberals, technocrats, and women are elected. In the *shura* councils of the other GCC states, rulers at least *appoint* technocrats and women. Two underlying conditions may be at work here: Firstly, the ruling families are truly interested in the development of their economies. This does not preclude that they enrich themselves but in several regards, they are more liberal and modern than their "subjects". Secondly, the Kuwaiti government partly lost control over the economic reform process and has to cope with often destructive powers. Instead of providing the government with necessary information on private sector necessities, parliament is used for rent-seeking and even sectarian quarrels.

The Kuwaiti government is somewhat restricted in its policies and this may also account for the rather lower ratings for government effectiveness and regulatory quality. Even if these ratings are higher than for Saudi Arabia, they are rather low compared to the other small GCC states.

In general, good growth performance only took place in the extremely autocratic countries and those with a sound government effectiveness and regulatory quality. In the context of the GCC states, government effectiveness and regulatory quality are of special importance, because development is initiated from above in an attempt to reach unbalanced growth after investment into the leading sector (oil and derivatives) and infrastructure. However, ~PARLIAMENT did not fully meet the 0.9 consistency for a necessary condition and it remains to be seen whether this condition plays a similar role in the test for sufficient conditions.

It is also of interest to note that political stability did not turn out to be a necessary condition. Looking at the ratings, Bahrain, Kuwait and Saudi Arabia display rather modest, and in the case of Bahrain, low political stability ratings. Nevertheless, Bahrain had good growth rates. Rather than assuming that political stability is of no importance for growth, I would argue that there are different kinds of political instability that are subsumed under the same figures. In Saudi Arabia, the perceived danger of terrorism threatens foreigners and foreign institutions. In Kuwait, the bad ratings were probably due to the Iraqi invasion and regional instability. In Bahrain, instability seems to be an intra-Bahraini thing: There are the fights for more participation and for Shiite rights. The perception of instability may be influenced by military repression. Even if it may sound cynical, it may be of economic advantage that the Bahraini government does not appease all conflicts by rent distribution and also risk unpopular measures such as the imposition of an income tax. There might be a rather counterintuitive link between economic inefficiency and political stability. Most GCC governments put a priority on stability be it at the expense of economic efficiency.

Reasons for this are the various gratifications and privileges for their nationals, which lead to ill-conceived incentives. Depending on the size of the national population, these will carry more or less weight. Hence, there may be the apparently paradoxical phenomenon of political stability being linked to rather bad diversification.

The resource abundance (RENT, RESERVES) and the dependence condition SXP also fail to be necessary. From this first analysis, one can conclude that failing to grow does not presuppose high rents or a high dependence on rents in exports. However, there are doubts about these results, as I will demonstrate below.

First Test for Sufficient Conditions

Given the limited number of cases, it is important to limit the conditions to those important for the special theoretical background and those which have not been found necessary for the outcome in question (NDGROWTH or ~NDGROWTH). In this case, these are: Investment behavior (INVESTCAPITA), political stability (POLSTAB), and a meaningful parliament (PARLIAMENT), as well as the different resource variables. Governance and regulatory quality (GOVREGULAR) will only be added to the test for ~NDGROWTH because it turned out to be necessary for a good diversification success (NDGROWTH). As such, it is always part of the configuration of sufficient conditions.

Again, I will test basic models first: INVESTCAPITA, POLSTAB, PARLIAMENT for NDGROWTH and INVESTCAPITA, POLSTAB, PARLIAMENT and GOVREGULAR for ~NDGROWTH. For the positive outcome, GOVREGULAR can be dropped because it is a necessary condition. For the negative outcome, Saudi Arabia is only covered by the solutions if GOVREGULAR is added as fsQCA output 2 shows. After testing this basic model, I will add the resource conditions separately. Solutions will be accepted if the consistency is higher than 0.85. The ultimate goal, however, will be to reach higher consistencies.

FsQCA Output 1 Basic Model for NDGROWTH

NDGROWTH = f(INVESTCAPITA, POLSTAB, PARLIAMENT) *Intermediate solution*

Frequency cutoff: 1.0; consistency cutoff: 0.818182

	Raw coverage	Unique coverage	Consistency
~PARLIAMENT*INVESTCAPITA	0.820690	0.820690	0.865455

Solution coverage: 0.820690

Solution consistency: 0.865455

*Cases with greater than 0.5 membership in term ~parliament*investcapita: UAE (0.83,1), Oman (0.82,0.8), Bahrain (0.65,1)*

FsQCA Output 2 Basic Model for ~NDGROWTH

~NDGROWTH = f(INVESTCAPITA, PARLIAMENT, POLSTAB, GOVREGULAR)	*Complex solution*		
Frequency cutoff: 1.0; consistency cutoff: 0.914894			
	Raw coverage	**Unique coverage**	**Consistency**
~INVESTCAPITA*POLSTAB*GOVREGULAR	0.570968	0.332258	0.956757
~INVESTCAPITA*~PARLIAMENT*~POLSTAB*~GOVREGULAR	0.325806	0.087097	0.926605

Solution coverage: 0.658065

Solution consistency: 0.962264

*Cases with greater than 0.5 membership in term ~investcapita*polstab*govregular: Qatar (0.77,0.9), Kuwait (0.55,1); Cases with greater than 0.5 membership in term ~investcapita*~parliament*~polstab* ~govregular: Saudi Arabia (0.55,1)*

Given the results for the necessary conditions, these solutions are not astonishing. FsQCA outputs 1 and 2 can be subsumed as follows:

> ~PARLIAMENT*INVESTCAPITA → NDGROWTH (Bahrain, Oman, UAE)
> POLSTAB*~INVESTCAPITA*GOVREGULAR → ~NDGROWTH (Kuwait, Qatar)
> ~INVESTCAPITA*~PARLIAMENT*~POLSTAB*~GOVREGULAR → ~NDGROWTH (Saudi Arabia)

The absence of a meaningful parliament in combination with good investment behavior was sufficient for a good diversification result in Bahrain, Oman, and the UAE. In Kuwait and Qatar, political stability and good governance as well as regulatory quality were not enough for a good outcome, obviously because they were combined with bad investment behaviors. In Saudi Arabia, the absence of good investment behavior, low political stability as well as low institutional quality, correlate with a failure to diversify despite the absence of a meaningful parliament.

Of course, GOVREGULAR, as a necessary condition, would also be part of the sufficient solution to the positive outcome, as can be seen from fsQCA output 3:
Please note that ISLAM could be used interchangeably with GOVREGULAR. Saudi Arabia is only covered if at least one of these conditions is added to the model. The solutions for the negative outcome would then look as follows:

FsQCA Output 3 Adding GOVREGULAR to the Basic Model for NDGROWTH

NDGROWTH = f(INVESTCAPITA, GOVREGULAR, POLSTAB, PARLIAMENT) *Complex solution*

Frequency cutoff: 1.0; consistency cutoff: 0.818182

	Raw coverage	Unique coverage	Consistency
INVESTCAPITA*GOVREGULAR*~PARLIAMENT	0.820690	0.820690	0.865455

Solution coverage: 0.820690

Solution consistency: 0.865455

*Cases with greater than 0.5 membership in term investcapita*govregular*~parliament: UAE (0.83,1), Oman (0.82,0.8), Bahrain (0.65,1)*

POLSTAB*~INVESTCAPITA*~ISLAM → ~NDGROWTH (Kuwait, Qatar)
~INVESTCAPITA*~PARLIAMENT*~POLSTAB*ISLAM → ~NDGROWTH (Saudi Arabia)

Interpretation of the Basic Model

The importance of a good investment behavior (INVESTCAPITA) for a positive diversification result as well as regulatory quality and governance were to be expected from the analysis of necessary conditions. Interestingly, the negative influence of a meaningful parliament is also confirmed. The absence of a meaningful parliament failed to be a necessary condition, but it is part of the configuration of sufficient conditions.

The condition POLSTAB is probably the most interesting. In the case of Qatar and Kuwait, it is part of the complex sufficient solution for the negative outcome. Despite the good institutional environment, the diversification process was not very successful. The crucial ingredient of the "recipe" for a negative outcome seems to be rather bad investment behavior and this calls for further explanations.

Combined with a low regulatory quality, the absence of political stability can also lead to a negative outcome, as in the case of Saudi Arabia. Hence, political stability is part of a path-dependent process. Its presence or absence alone does not indicate whether NDGROWTH is absent or present. The influence of political stability on NDGROWTH seems to depend on GOVREGULAR. A government with high regulatory quality can "regulate" the economy and the welfare system in a way that makes political stability likely. It can "buy" political stability at the price of economic inefficiencies. Without regulatory quality, the reasons for political instability would probably become more obvious. This is one explanation, but only a very shallow one. It will have to be asked why some governments have the means to regulate their countries well and others do not.

As expected, INVESTCAPITA is part of all solutions: ~INVESTCAPITA is an INUS cause for ~NDGROWTH, while INVESTCAPITA is an INUS cause for NDGROWTH. This means that a rather bad investment behavior is part of the sufficient solution for a failure to diversify, while a good investment behavior is part of the sufficient solution for successful diversification. Looking at the parsimonious solutions, I found that INVESTCAPITA is always the parsimonious solution for NDGROWTH, while ~INVESTCAPITA is always the parsimonious solution for ~NDGROWTH. Hence, a good investment behavior is crucial but it remains to be seen why some countries invest most of their rents locally and others not. This seems to be a matter of choice for the governments and this choice requires further explanations.

Adding the Resource Conditions to the Model – is a Country Always a Case?

At this point, the most interesting conditions – the resource conditions – have yet to be added to the model. For the following fsQCA outputs, the complex solution is identical with the intermediary solution for NDGROWTH. Again, the first condition to be added is the annual average resource rent per capita (RENT). FsQCA output 4 summarizes the test for the positive outcome.

FsQCA Output 4 Adding RENT to the Test for NDGROWTH

NDGROWTH = f(INVESTCAPITA, POLSTAB, GOVREGULAR, PARLIAMENT, RENT) — *Complex solution*

Frequency cutoff: 1.0; consistency cutoff: 0.835165

	Raw coverage	**Unique coverage**	**Consistency**
INVESTCAPITA*GOVREGULAR*~PARLIAMENT*~RENT	0.486207	0.486207	0.903846

Solution coverage: 0.486207

Solution consistency: 0.903846

*Cases with greater than 0.5 membership in term investcapita*govregular*~parliament*~rent: Oman (0.8,0.8), Bahrain (0.61,1)*

The absence of a high annual *per capita* resource rent is confirmed as a part of the sufficient condition for a positive outcome, but the "case" of the UAE is lost, as fsQCA output 4 shows.

Testing for ~NDGROWTH, a high annual *per capita* resource rent is part of the sufficient solution for the failure to diversify, and again, Saudi Arabia is only covered by the analysis if GOVREGULAR is part of the model. (See fsQCA output 5)

FsQCA Output 5 Adding RENT to the Test for ~NDGROWTH

~NDGROWTH = f(INVESTCAPITA, GOVREGULAR, PARLIAMENT, POLSTAB, RENT) *Complex solution*

Frequency cutoff: 1.0; consistency cutoff: 0.907692

	Raw coverage	Unique coverage	Consistency
~INVESTCAPITA*GOVREGULAR*POLSTAB*RENT	0.570968	0.454839	0.956757
~INVESTCAPITA*~GOVREGULAR*~PARLIAMENT* ~POLSTAB*~RENT	0.190323	0.074194	0.907692

Solution coverage: 0.645161

Solution consistency: 0.961538

*Cases with greater than 0.5 membership in term ~investcapita*govregular*polstab*rent: Qatar (0.77,0.9), Kuwait (0.55, 1); Cases with greater than 0.5 membership in term ~investcapita*~govregular*~parliament*~polstab*~rent: Saudi Arabia (0.51,1)*

RENT behaves as expected and the resource curse seems to be confirmed. However, the small sample size of this analysis does not allow for a dropping of cases. Using another resource condition does not solve the problem: Adding resource reserves (RESERVES) or the resource exports as a share of GNP (SXP), the UAE is equally lost. If RENTIERII is added (which was a necessary condition for ~NDGROWTH), all cases are covered, but the consistency is only slightly above 0.8.

At this point, the UAE should be looked at more closely. In particular, their annual average resource rent and their reserves are of interest because the UAE was covered before these conditions were added.

The UAE is a loose federation without a common economic policy. Nevertheless, the emirates are usually treated as one case in regressions and, unfortunately, also by most macroeconomic data. In several aspects the emirates are similar, but they are highly heterogeneous with respect to their natural resource production and reserves, as table 44 demonstrates.

Table 44 Repartition of Oil and Gas Reserves and Production in the UAE

	Production Oil in mb/day	Reserves Oil in bbl	Production Gas in mcm	Reserves Gas in bcm
Abu Dhabi	1.990	92.2	14.3	6,003
Dubai	0.230	5.9	1.7	121
Sharjah	0.058	n.a.	0.20	303
Ras al-Khaimah	n.a.	n.a.	0.04	31

Sources: Butt 2001: 247-248.

Hence, if the emirates were treated as separate cases, they would probably differ the most with respect to the resource conditions. Early data for a single emirate is hard to obtain. From qualitative knowledge, however, it is possible to estimate data for at least the two most important emirates: Dubai and Abu Dhabi.

I assume that Abu Dhabi is roughly comparable to Qatar rather than to Dubai. While Dubai and Abu Dhabi are comparable with respect to their level of autocracy, only Abu Dhabi has vast reserves and is immensely resource rich in *per capita* terms. Rents should also contribute the lion's share to government revenue, making it a high level rentier. The natural resource exports should also make up a high percentage of GNP, given that Abu Dhabi started diversification significantly later than Dubai.[1] The institutional ratings should be rather sound: Abu Dhabi is politically stable, probably even more stable than Dubai, and certainly more stable than Sharjah. More than Dubai, Abu Dhabi tries to uphold Islamic values. This is true for daily life, alcohol, and nightclubs, but also for the legal system where *sharia* courts still play a bigger role. (al-Muhairi 1996) Universities are separated into male and female campuses (see i.e., Zayed University, Higher Colleges for Technology, etc.) Most tourists still travel to Dubai and not to Abu Dhabi if they go to the UAE.

Experts agree that Abu Dhabi tackled diversification much more slowly and probably also more prudently than Dubai. Recently, there have been a few daring projects and more pronounced diversification efforts. Still, Abu Dhabi's overall non-oil growth can be expected to be rather modest and not disastrous.

Dubai, on the other hand, has had the highest non-oil growth rates even if the sustainability of the "Dubai model" remains questionable. Dubai's reserves and rents are rather modest: Many projects were financed by debts, leading to a debt crisis in the context of the international financial and economic crisis of 2008. (Evans-Pritchard 2008) Numbers for "rentier" can only be estimated, but some sources estimate around 30 to 40 percent of government revenue in times of modest oil prices. To increase its income, Dubai has already introduced several indirect taxes.[2] Dubai diversified rather early and has always been an important trading hub. The share of natural resource exports in GNP should therefore be lower than elsewhere. Using these estimates, the table of fuzzy values could look as in table 45.

Testing the New Sample of Seven Cases

The sample now consists of seven cases (the UAE as a federation will be dropped). I tested again for necessary conditions. Table 46 shows that the necessary conditions have not changed, only slightly in consistency and coverage. None of the resource abundance conditions turned out to be necessary. Only ~RENTIERII remained necessary for ~NDGROWTH, but at a low consistency of 0.92. Again, I will postpone the explanation of this result until the behavior of RENTIERII as a sufficient condition is clear.

Table 45 Fuzzy Sets for GCC States (7 cases)

	NDGROWTH	INVESTCAPITA	PARLIAMENT	POLSTAB	RULEOFLAW	GOVREGULAR
Abu Dhabi	0.2	0.25	0.17	0.9	0.8	0.95
Bahrain	1	0.65	0.17	0.06	0.88	0.94
Dubai	1	1	0.17	0.8	0.99	0.95
Kuwait	0	0.08	0.59	0.55	0.95	0.71
Oman	0.8	0.83	0.18	0.85	0.93	0.93
Qatar	0.1	0.23	0.09	0.9	0.89	0.86
Saudi Arabia	0	0.14	0.05	0.45	0.69	0.28
UAE	1	0.98	0.17	0.91	0.95	0.95
	ISLAM	RENT	RESERVES	SXP	RENTIERI	RENTIERII
Abu Dhabi	0.2	1	1	0.95	0.97	0.86
Bahrain	0	0.39	0.23	0.18	0.62	0.25
Dubai	0	0.35	0.3	0.2	0	0.2
Kuwait	0	0.99	0.97	0.88	0.98	0.94
Oman	0.2	0.2	0.4	0.99	0.96	0.84
Qatar	0.1	1	1	0.95	0.98	0.93
Saudi Arabia	1	0.49	0.43	0.6	0.95	0.82
UAE	0.2	1	0.9	0.9	0.97	0.89

Note: The UAE as a federation was excluded.

Table 46 Testing for Necessary Conditions (7 cases)

Conditions Tested	For NDGROWTH		For ~NDGROWTH	
	Consistency	Coverage	Consistency	Coverage
INVESTCAPITA	*0.89*	0.86	0.21	0.34
~INVESTCAPITA	0.26	0.21	*0.88*	0.89
POLSTAB	0.63	0.43	0.68	0.75
~POLSTAB	0.48	0.6	0.4	0.57
RULEOFLAW	**0.96**	0.48	*0.88*	0.63
~RULEOFLAW	0.16	0.57	0.17	0.82
GOVREGULAR	0.96	0.53	0.66	0.55
~GOVREGULAR	0.11	0.24	0.39	0.95
ISLAM	0.16	0.33	0.42	1
~ISLAM	**1**	0.56	0.68	0.57
PARLIAMENT	0.25	0.55	0.29	0.84
~PARLIAMENT	*0.89*	0.49	0.79	0.63
RENT	0.4	0.28	0.83	0.84
~RENT	0.66	0.8	0.23	0.37
RESERVES	0.4	0.28	0.81	0.83
~RESERVES	0.67	0.78	0.26	0.41
SXP	0.48	0.31	0.83	0.72
~SXP	0.56	0.77	0.19	0.41
RENTIERII	0.5	0.32	**0.92**	0.76
~RENTIERII	0.62	0.89	0.15	0.39

Testing the same basic model as before, all states are covered. Again, Saudi Arabia is only covered if the condition on governance and regulatory quality (GOVREGULAR) is included. The solutions are basically the same as for the sample of six cases:

Complex solution for INVESTCAPITA, POLSTAB and PARLIAMENT, tested for NDGROWTH
INVESTCAPITA*~PARLIAMENT → NDGROWTH
(consistency: 0.86)

Complex solutions for INVESTCAPITA, POLSTAB, GOVREGULAR and PARLIAMENT, tested for ~NDGROWTH
~INVESTCAPITA*POLSTAB*GOVREGULAR → ~NDGROWTH (Qatar, Abu Dhabi, Kuwait)
~INVESTCAPITA*~POLSTAB*~GOVREGULAR*~PARLIAMENT → ~NDGROWTH (Saudi Arabia)
(consistency 0.98)

Without GOVREGULAR, the solution would not cover Saudi Arabia and would just look as follows:
~INVESTCAPITA*POLSTAB → ~NDGROWTH (parsimonious, intermediate and complex solution)

Adding the Annual Average Resource Rent per capita

If RENT is added to the model, both Dubai and Abu Dhabi are explained. Dubai is explained by the solution for the positive outcome in fsQCA output 6, while Abu Dhabi can be found in fsQCA output 7.

FsQCA Output 6 Adding RENT to the Test for NDGROWTH, seven cases

NDGROWTH = f(INVESTCAPITA, POLSTAB, PARLIAMENT, RENT) *Complex solution*

Frequency cutoff: 1.0; consistency cutoff: 0.864865

	Raw coverage	Unique coverage	Consistency
INVESTCAPITA*~PARLIAMENT*~RENT	0.664516	0.664516	0.932127

Solution coverage: 0.664516

Solution consistency: 0.932127

Cases with greater than 0.5 membership in term investcapita~parliament*~rent: Oman (0.8,0.8), Dubai (0.65,1), Bahrain (0.61,1)*

FsQCA Output 7 Adding rent to the Test for ~ndgrowth, seven cases

~NDGROWTH = f(INVESTCAPITA, POLSTAB, PARLIAMENT, GOVREGULAR, RENT)			*Intermediate solution*
Frequency cutoff: 1.0; consistency cutoff: 0.907692			
	Raw coverage	**Unique coverage**	**Consistency**
RENT*GOVREGULAR*POLSTAB*~INVESTCAPITA	0.646154	0.553846	0.976744
~RENT*~GOVREGULAR*~PARLIAMENT*~POLSTAB*~INVESTCAPITA	0.151282	0.058974	0.907692

Solution coverage: 0.705128

Solution consistency: 0.978648

*Cases with greater than 0.5 membership in term rent*govregular*polstab*~investcapita: Qatar (0.77,0.9),*
Abu Dhabi (0.75,0.8), Kuwait (0.55,1)
Cases with greater than 0.5 membership in term ~rent~govregular*~parliament*~polstab*~investca pita:*
Saudi Arabia (0.51,1)

For the parsimonious solution, the computer suggests, as suspected above, two simple solutions:

INVESTCAPITA → NDGROWTH
~INVESTCAPITA → ~NDGROWTH

Adding Resource Reserves

Basically the same happens when RESERVES is added to the model, which is not astonishing because of the covariance of the two conditions. The summarized results can be taken from tables 50 and 51 below.

Adding the Share of Natural Resource Exports in GNP (SXP)

Adding SXP leads to a lower consistency and the case of Oman is unexplained as fsQCA outputs 8 and 9 show:

FsQCA Output 8 Adding SXP to the Test for NDGROWTH, seven cases

NDGROWTH = f(INVESTCAPITA, POLSTAB, PARLIAMENT, SXP) *Complex solution*
Frequency cutoff: 1.0; consistency cutoff: 0.813559

	Raw coverage	Unique coverage	Consistency
INVESTCAPITA*~PARLIAMENT*~SXP	0.503226	0.503226	0.876404

Solution coverage: 0.503226
Solution consistency: 0.876404
Cases with greater than 0.5 membership in term investcapita~parliament*~sxp: Dubai (0.8,1), Bahrain (0.65,1)*

On the other hand, the test for a failure to diversify (~NDGROWTH) seems to explain Saudi Arabia, even without adding GOVREGULAR.

FsQCA Output 9 Adding SXP to the Test for ~NDGROWTH, seven cases

~NDGROWTH = f(INVESTCAPITA, POLSTAB, PARLIAMENT, SXP) *Complex solution*
Frequency cutoff: 1.0; consistency cutoff: 0.879195

	Raw coverage	Unique coverage	Consistency
~INVESTCAPITA*~PARLIAMENT*SXP	0.692308	0.038462	0.937500
~INVESTCAPITA*POLSTAB*SXP	0.689744	0.035897	0.978182

Solution coverage: 0.728205
Solution consistency: 0.940397

Cases with greater than 0.5 membership in term ~investcapita~parliament*sxp: Qatar (0.77,0.9), Abu Dhabi (0.75,0.8), Saudi Arabia (0.6,1)*

*Cases with greater than 0.5 membership in term ~investcapita*polstab*sxp: Qatar (0.77,0.9), Abu Dhabi (0.75,0.8), Kuwait (0.55,1)*

The same occurs when adding the condition of rents as a percentage of government income (RENTIERII) as fsQCA outputs 10 and 11 demonstrate:

FsQCA Output 10 Adding RENTIERII to the Test for NDGROWTH, seven cases

NDGROWTH = f(INVESTCAPITA, POLSTAB, PARLIAMENT, RENTIERII) *Complex solution*

Frequency cutoff: 1.0; consistency cutoff: 0.854014

	Raw coverage	Unique coverage	Consistency
INVESTCAPITA*~PARLIAMENT*~RENTIERII	0.587097	0.587097	0.900990

Solution coverage: 0.587097
Solution consistency: 0.900990

Cases with greater than 0.5 membership in term investcapita~parliament*~rentierii: Dubai (0.8,1), Bahrain (0.65,1)*

FsQCA Output 11 Adding RENTIERII to the Test for ~NDGROWTH, seven cases

~NDGROWTH = f(INVESTCAPITA, POLSTAB, PARLIAMENT, RENTIERII) *Complex solution*

Frequency cutoff: 1.0; consistency cutoff: 0.839744

	Raw coverage	Unique coverage	Consistency
~INVESTCAPITA*~PARLIAMENT*RENTIERII	0.748718	0.094872	0.921136
~INVESTCAPITA*POLSTAB*RENTIERII	0.689744	0.035897	0.978182

Solution coverage: 0.784615

Solution consistency: 0.924471

Cases with greater than 0.5 membership in term ~investcapita~parliament*rentierii: Saudi Arabia (0.82,1), Qatar (0.77,0.9), Abu Dhabi (0.75,0.8)*

*Cases with greater than 0.5 membership in term ~investcapita*polstab*rentierii: Qatar (0.77,0.9), Abu Dhabi (0.75,0.8), Kuwait (0.55,1)*

Summary of the Solutions

These different models have produced a whole array of solutions, which can be found summarized in the following tables. Table 47 repeats the necessary solutions, while tables 48 and 49 summarize the sufficient solutions' terms for the positive and negative outcome.

Table 47 Results – Necessary Conditions (7 cases)

Condition	Consistency	Necessary for
GOVREGULAR	0.96	NDGROWTH
~ISLAM	1	NDGROWTH
RENTIERII	0.92	~NDGROWTH

Table 48 Basic Model with Resource Conditions for NDGROWTH (complex/intermediate solutions)

	→NDGROWTH	Cases Covered
Basic Model	INVESTCAPITA*~PARLIAMENT → NDGROWTH Solution consistency: 0.86	Bahrain, Dubai, Oman
RENT	INVESTCAPITA*~PARLIAMENT*~RENT Solution consistency: 0.93	Bahrain, Dubai, Oman
RESERVES	INVESTCAPITA*~PARLIAMENT*~RESERVES Solution consistency: 0.92	Bahrain, Dubai, Oman
SXP	INVESTCAPITA*~PARLIAMENT*~SXP Solution consistency: 0.88	Bahrain, Dubai
RENTIERII	INVESTCAPITA*~PARLIAMENT*~RENTIERII Solution consistency: 0.9	Bahrain, Dubai

Table 49 Basic Model with Resource Conditions for ~NDGROWTH (complex/intermediate solutions)

	~NDGROWTH	Cases Covered
Basic Model	~INVESTCAPITA*POLSTAB*GOVREGULAR → ~NDGROWTH	Abu Dhabi, Kuwait, Qatar
	~INVESTCAPITA*~POLSTAB*~GOVREGULAR*~PARLIAMENT → ~NDGROWTH	Saudi Arabia
	Solution consistency: 0.98	
RENT	RENT*GOVREGULAR*POLSTAB*~INVESTCAPITA	Abu Dhabi, Kuwait, Qatar
	~RENT*~GOVREGULAR*~PARLIAMENT*~POLSTAB*~INVESTCAPITA	Saudi Arabia
	Solution consistency: 0.98	
RESERVES	~INVESTCAPITA*POLSTAB*GOVREGULAR*RESERVES	Abu Dhabi, Kuwait, Qatar
	~INVESTCAPITA*~POLSTAB*~GOVREGULAR*~PARLIAMENT*~RESERVES	Saudi Arabia
	Solution consistency: 0.98	
SXP	~INVESTCAPITA*~PARLIAMENT*SXP	Qatar, Abu Dhabi, Saudi Arabia
	~INVESTCAPITA*POLSTAB*SXP	Qatar, Abu Dhabi, Kuwait
	Solution consistency: 0.94	
RENTIERII	~INVESTCAPITA*~PARLIAMENT*RENTIERII	Qatar, Abu Dhabi, Saudi Arabia
	~INVESTCAPITA*POLSTAB*RENTIERII	Qatar, Abu Dhabi, Kuwait
	Solution consistency: 0.92	

Interpretation

Obviously, the changed sample has not had any influence on the necessary conditions or on the results for the basic model. This was to be expected: When "splitting up" the UAE into Abu Dhabi and Dubai, only the "resource conditions" differed considerably. Hence, there is no need to interpret the basic model again. The models including the resource conditions, however, call for an explanation. In general, the consistency of sufficient solutions was higher than 0.9 and most models covered all cases. Considering the restrictions which stem from the small sample size and the resulting limited variance, the results are rather solid.

The first remarkable finding is the clear difference between the resource dependence and the resource abundance measures. The two resource abundance measures (RENT and RESERVES) produce solutions that cover all cases. The absence of high rents or high reserves is part of the sufficient path to successful diversification. Low rents or reserves (*per capita*) are not in themselves sufficient because in Saudi Arabia, in combination with low ratings for governance and regulatory quality (~GOVREGULAR), even low *per capita* rents are not sufficient to produce a positive outcome. Nevertheless, three out of four countries that failed to have a positive outcome have high rents and reserves *and* good institutions.

Of course, Dutch disease could be at work here but high rents in the GCC also correlate with abundant reserves of oil and gas, and the reason for lower growth could be found in fewer incentives to diversify: There are ample resources to sustain the standard of living for the future generations also, hence there is no hurry to diversify and risk unpopular economic policies. This could also be reflected in the lower INVESTCAPITA values: There is no need to invest the rents locally. Here, a glance at theory and case studies can help to decide. As mentioned in the theory chapter, Dutch disease effects are mitigated by strong government involvement, migrant workers, and planned investments from above. Furthermore, the governments are well aware that measures to avoid currency appreciation had to be taken. While Dutch disease effects may have taken place, it is rather unlikely that these are the only explanations for growth problems. Furthermore, qualitative studies also often highlight the incentives to diversify, as in the case of Qatar, which plans for a future *with* oil and gas and not without. (Hvidt 2013: 27)

It should also be noted that the "incentives" effect is not only mirrored by RESERVES, but also by RENT. When large amounts of rents have poured into a country for decades and have been used to privilege the nationals, there is a habitation effect. No matter if the resources will be depleted soon or not, people will try to uphold their privileges and standard of living. The scenario of a population renouncing its privileges voluntarily because the upcoming depletion of resources dictates it is rather unlikely.

The resource *dependence* measures, frequently significant in cross-country regressions, explain much less and even lead to contradictions. Low RENTIERII or SXP values have only been reached by countries that have started to diversify early *and* have relatively low rents, such as the case of Bahrain and Dubai.[3] There are two ways to reach high dependency levels: In the first case, a country has very low non-oil GDP and very low rents and reserves are enough to make it dependent on its rents. This is the case of Oman, especially at the beginning of the 1970s. Testing SXP and RENTIERII for NDGROWTH, Oman is not covered. In the case of Oman, high SXP and RENTIERII values in combination with low reserves and rents stand for high incentives to overcome this dependence. Other countries with extremely high SXP and RENTIERII values, such as Kuwait and Qatar have ample reserves and rents that decrease their incentives to diversify rapidly. Thus, similar SXP and RENTIERII values stand for entirely different situations, which, in turn, are the cause for entirely different incentives (see table 50). It seems that resource dependence measures such as the share of natural resource exports in GNP "infiltrated" empirical studies on natural resources and growth without a check against theory.

Table 50 Heterogenous Backgrounds of SXP and RENTIERII

High SXP and high RENTIERII, but rather low RENT and RESERVES (*per capita*)	Oman, to a lesser extent Saudi Arabia
High SXP and high RENTIERII ***and*** high RENT and RESERVES	Qatar, Kuwait, Abu Dhabi

Rentierism and the underlying concept of measuring the percentage of rents in government income help to define the rentier state, which in turn is associated with certain characteristics. When it comes to economic inefficiencies, rentier state theory argues on the basis of incentives, corruption, rent seeking, and rentier mentality, and so the conditions RENT and RESERVES seem more appropriate to reflect these mechanisms. A high share of rents in government income, on the other hand, can be regarded as a sign of low taxation or no taxation at all. This absence of taxes, in turn, is thought to explain why rentier states are mostly autocratic states ("No representation without taxation"). A potential mechanism triggered by a high share of rents in government revenue seems to be based on the political consequences of resource abundance rather than on its impact on economic growth. To be blunt: There is no reason why SXP or RENTIERII should explain a good or bad growth performance.

Contrary to the regression, the fsQCA analysis reveals that RENTIERII and SXP reflect heterogeneous situations by dropping cases. Oman is not explained, while Saudi Arabia seems to be explained by its high dependence on natural resource rents, reflected in high SXP and RENTIERII values. What high SXP and RENTIER values, however, can stand for is a long-standing inability to overcome dependence, especially if non-resource GDP or exports fail to grow or because there are no taxes? If high dependence is already reached, this may point to poverty traps and other, often self-perpetuating weaknesses, and the importance of these variables/conditions can be a self-fulfilling prophecy. Having been unable to overcome dependence on resources in the past there may be persistent inhibitions to growth.

Hence, the validity of RENTIERII as a necessary condition for ~NDGROWTH has to also be interpreted in this light. Necessity means that all negative performers show high RENTIERII values, but high RENTIERII is not sufficient to explain a bad performance – as the case of Oman shows. However, those countries with high RENTIERII that are bad performers are identical with those that have high rents or high reserves: Qatar, Kuwait, and Abu Dhabi. Saudi Arabia is at the turning point of having high rents and reserves but the determining factor here are its bad institutions. It seems that the choice of the sample has added a lot to the fact that RENTIERII has turned out to be necessary.

Methodological Interpretation

Already after this first QCA-Analysis, several advantages can be determined. The concentration on a few cases allows for the questioning of widespread operationalization practices of regression variables. In more than one case, it became obvious that these operationalizations stem from the context of countries where natural resources play only modest roles for the economy. This is not a new finding but small samples allow for at least some adaptation.

Regressions in general – and panel analyses in particular – have to rely on good data quality. Data for developing countries, however, are often of bad quality. Moreover, the GCC states, due to their relatively good financial situation, are not closely supervised by the IMF, and there is therefore no necessity to provide good data. This is reflected in the PENN World Table grades. On a scale from A data (sound quality) to D data (low quality), Bahrain, Kuwait, Oman and Qatar received a "C", whereas Saudi Arabia and the UAE reached only a "D".[4]

In such a case, it is an advantage to work with fuzzy sets. They do not only allow for a simplification of data which looks too detailed to be true, but also permit a comparison of the data with qualitative studies, and insight from the latter can change the calibration.

The fuzzy set calibration was also an advantage when the UAE had to be split up. Fuzzy sets can display rather vague assessments better than a regression: For a regression, seemingly precise data would have to be invented.

It is, however, unlikely that the UAE would have attracted the attention of the researcher in a large sample. The fact that the UAE may be a country but not a case became obvious when looking closely at the sample, and also at the methodology that highlights contradictions instead of averaging results. In a regression sample, such a close look would probably go beyond the scope of the analysis, even if there are certainly more cases like the UAE.

Furthermore, the sample choice may have been an advantage. Comparing very resource rich countries to less resource rich countries puts the focus on the difference between extremely resource rich countries that can live on their resources for generations to come, and those that are only resource rich but cannot afford to remain idle. A regression (as long as no dummy variable is used for the natural resources) assumes a proportional relationship: The more resources, the lower the growth rate. However, there may be a turning point: Up to a certain degree of resource abundance, natural resources may be an asset that can contribute to development. At such levels of resource abundance, no one can reasonably assume to live on the rents from natural resource exports. The incentives to develop are as high as for any other country; maybe even higher because the high standard of living attained due to natural resource wealth has to be maintained. At a certain degree of resource abundance, however, it becomes possible to live on natural resource rents, and incentives to diversify may drop sharply.

The small sample also had several flaws, mostly due to a lack of variance. Some conditions could not be used as intended because membership scores of the sample countries were too similar. In other cases, there was only one case which deviated significantly from the rest of the sample and this is, admittedly, a weak ground for general assumptions.

5.3 Enlargement of the Sample: The GCC States within a Sample of Oil Exporters

An entirely different perspective can be gained from an enlarged sample. The GCC states are extreme examples for resource abundance, and display a certain homogeneity regarding cultural and historical patterns. The following analysis puts the GCC in comparison to other natural resource exporters. The main criterion was the presence of oil or of considerable wealth in other resources (diamonds in Botswana, for example).

Due to the heterogeneous character of this sample, several conditions have to be changed, or rather, will be used in their "normal" version as indicated in the methodological chapter. Since not all countries in this sample are resource dependent, the common measures – such as the investment share of GDP – are valid and adapting these measures to the specific situation of extremely resource dependent countries like the GCC states would make little sense. I have thus opted for the investment share of GDP (INVESTMENT) instead of the investments relative to the rentier income (INVESTCAPITA), a compromise that the interpretation has to take into account. Information on women in the labor force and fertility rates can no longer be included in the condition on Islamic economic institutions because now the sample is more heterogeneous. Low participation rates and high fertility rates can have entirely different reasons in non-Muslim countries. These two conditions are now added separately. (FERTILITY and WOMENLAB)

Moreover, limited variance is unlikely to occur in such a heterogeneous sample. Conditions that could not be interpreted in a meaningful way due to covariance are not problematic anymore: RULEOFLAW can be used as a valid condition and DEMOCRACY can be tested without having to recur to PARLIAMENT as above. There will be a condition on human capital based on the average years of schooling (HUMANCAP). Instead of RENTIERII, RENTIERI will be used, which divides the sample into rentiers and semi-rentiers (or non-rentiers). Using RENTIERI, all GCC countries will now have fuzzy values higher than 0.5. For the calibration of the resource abundance conditions, the turning point is still between the extremely resource rich and the *less* resource rich rather than the resource poor.

As a learning-effect from the previous analysis, the UAE appears in the data-set as Abu Dhabi and Dubai, with the estimated data from above.

Necessary Conditions

The heterogeneity of this sample reveals itself after the first steps of the analysis: As table 51 shows, only two conditions turn out to be necessary, while one of them is not very consistent (0.91). Only two conditions come close to necessity, and one of them is a resource conditions.

Table 51 **Test for Necessary Conditions, enlarged sample**

	Tested for NDGROWTH		Tested for ~NDGROWTH	
	Consistency	coverage	Consistency	Coverage
NDINITIAL	0.51	0.58	0.54	0.59
~NDINITIAL	0.63	0.59	0.61	0.54
INVESTMENT	0.73	0.6	0.74	0.58
~INVESTMENT	0.48	0.66	0.48	0.63
HUMANCAP	0.53	0.64	0.49	0.56
~HUMANCAP	0.63	0.56	0.68	0.58
DEMOCRACY	0.56	0.67	0.45	0.51
~DEMOCRACY	0.59	0.53	0.71	0.6
POLSTAB	0.64	0.64	0.5	0.48
~POLSTAB	0.47	0.5	0.62	0.62
RULEOFLAW	0.71	0.63	0.53	0.45
~RULEOFLAW	0.39	0.47	0.57	0.65
GOVREGULAR	0.75	0.64	0.56	0.45
~GOVREGULAR	0.36	0.46	0.56	0.68
WOMENLAB	0.62	0.65	0.48	0.48
~WOMENLAB	0.51	0.5	0.65	0.62
FERTILITY	0.69	0.49	*0.85*	0.57
~FERTILITY	0.38	0.72	0.23	0.42
RENTIERI	0.41	0.46	0.66	0.7
~RENTIERI	0.73	0.69	0.49	0.44
SXP	0.16	0.34	0.38	0.75
~SXP	*0.88*	0.6	0.67	0.43
RENT	0.18	0.37	0.42	0.81
~RENT	**0.91**	0.62	0.68	0.44
RESERVES	0.42	0.52	0.65	0.77
~RESERVES	*0.82*	0.71	0.6	0.5
ARAB	0.35	0.46	0.43	0.54
~ARAB	0.65	0.54	*0.57*	0.46
PERCENTMUSLIM	0.46	0.52	0.51	0.55
~PERCENTMUSLIM	0.6	0.56	0.56	0.5
ISLAM	0.12	0.46	0.24	0.86
~ISLAM	**0.96**	0.57	0.85	0.48

As in the first sample, ~ISLAM is a necessary condition for NDGROWTH at a consistency of 0.96. While the consistency is acceptable, it should also be noted that there are only two cases with a fuzzy value for ISLAM higher than 0.5. Apart from Saudi Arabia, only Iran is such a case; hence there is only little variance in the sample on this issue. Nevertheless the two countries with ISLAM higher than 0.5 both are not among the good performers. At 0.91, the absence of high rents, and also, at 0.88, the absence of a high SXP (share of resource exports in GDP as of 1970) come closest to necessary conditions. The absence of high reserves only reaches 0.82 as a necessary condition for a sound diversification success (NDGROWTH). This gap between RENT and RESERVES points to the fact that the two conditions are no longer covariant in this sample.

Consistency, however, is rather low in all cases and there are still contradicting cases, even for the consistency of ~RENT, as the following plot shows. The contradicting cases are mostly GCC states: Abu Dhabi, Bahrain, Dubai, Qatar but also Botswana and Indonesia or Malaysia, depending on the condition. Apparently, their annual average resource rents are still too high if compared to their diversification success: At such a level of resource abundance, they should not have attained high NDGROWTH values. However, they are all fairly close to the diagonal and not located in the corner as the fsQCA plot (figure 6) shows:

Figure 6 **FsQCA Plot ~RENT as a Necessary Condition**

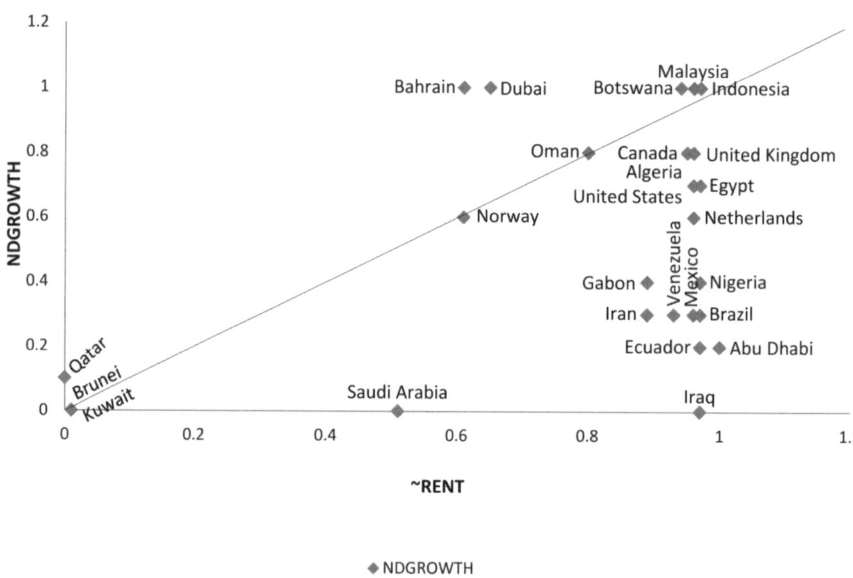

A few other conditions have consistency values over 0.7. Not surprisingly, these are those conditions that were necessary in the smaller sample: INVESTMENT and GOVREGULAR, while the lower consistency points to exceptions.

Sufficient Conditions

The inclusion of further cases allows for a test with more conditions. As in the regression, I will first test a basic model: INVESTMENT, POLSTAB, one institutional condition (RULEOFLAW or GOVREGULAR) and one condition on women (FERTILITY or WOMENLAB). As a single condition, none of them is sufficient. The consistency threshold of a sufficient solution will be set at 0.85.

The first models to be tested for both good and bad diversification results (NDGROWTH and ~NDGROWTH) look as follows:

a, INVESTMENT, POLSTAB, RULEOFLAW and FERTILITY
b, INVESTMENT, POLSTAB, RULEOFLAW and WOMENLAB
c, INVESTMENT, POLSTAB, GOVREGULAR and FERTILITY
d, INVESTMENT, POLSTAB, GOVREGULAR and WOMENLAB

Also HUMANCAP will be added instead of FERTILITY/WOMENLAB:

e, INVESTMENT, POLSTAB, RULEOFLAW and HUMANCAP
f, INVESTMENT, POLSTAB, GOVREGULAR and HUMANCAP
as well as the two conditions on Arab society and Islam:
G, INVESTMENT, POLSTAB, RULEOFLAW and ARAB
h, INVESTMENT, POLSTAB, GOVREGULAR and ARAB
I, INVESTMENT, POLSTAB, GOVREGULAR and PERCENTMUSLIM
J, INVESTMENT, POLSTAB, GOVREGULAR and PERCENTMUSLIM

Tested for both NDGROWTH and ~NDGROWTH, these configurations of conditions cover only few cases as table 52 shows. Lowering the consistency threshold to 0.8, coverage rises but the analysis far from covers the sample, as the following table shows. When testing for the positive outcome, only one or two cases can be assigned to configurations with a consistency higher than 0.8. Testing for the negative outcome, more cases are covered but still not enough for a reasonable analysis. Furthermore, even if those few cases covered, the consistencies are rather low. The results are not improved by testing combinations of all five conditions used above.

Table 52 Basic Models – Coverage, enlarged sample

Model Tested	For NDGROWTH Cases covered	For ~NDGROWTH Cases covered
INVESTMENT, POLSTAB, RULEOFLAW, and FERTILITY	2	4
INVESTMENT, POLSTAB, RULEOFLAW, and WOMENLAB	2	4
INVESTMENT, POLSTAB, GOVREGULAR, and FERTILITY	1	5
INVESTMENT, POLSTAB, GOVREGULAR, and WOMENLAB	1	6
INVESTMENT, POLSTAB, RULEOFLAW, and HUMANCAP	2	4
INVESTMENT, POLSTAB, GOVREGULAR, and HUMANCAP	1	5

Using fewer conditions is also not conducive: Testing, for instance, only investments (INVESTMENT) and political stability (POLSTAB), not a single country is explained.

The low coverage for testing ~NDGROWTH, however, should not be too surprising: Many cases are implicitly explained by the almost-necessary condition for NDGROWTH, which is ~RENT. Cases with high RENT, if they are low performers, are then explained. RENT, however, is still absent from this model.

The best consistencies can be obtained by constellation d: INVESTMENT, POLSTAB, GOVREGULAR, and WOMENLAB. GOVREGULAR turned out to be the most significant necessary condition in the previous analysis and performed better than RULEOFLAW. Looking at the data of WOMENLAB and FERTILITY there is an expected relationship: The high percentage of women in the labor force goes hand in hand with a lower birth rate. The exception are African states and to a certain extent, Malaysia. Here, high birth rates and a high percentages of women in the work force can be found – even if this may be the informal labor market: Women often work in agriculture or run small businesses to feed their family. They are reported to work twice as long as men, but their number in wage employment lags behind these apparently high numbers, especially in Nigeria. (IFAD 1999) Thus, the relatively high percentages of women in the labor force do not necessarily correlate with more gender equality, or access of women to education and vocational training. Nevertheless, agriculture and trade are important and women's contribution to growth seems to be reflected more in female labor force participation rates than in fertility rates. For this stage of the analysis, the use of WOMENLAB and GOVREGULAR can be justified. Nevertheless, it should be noted that the aspect of population growth, which is important due to the use of *per capita* growth rates, is not necessarily reflected in WOMENLAB. As a result, my first aggregate model includes the following conditions: INVESTMENT, POLSTAB, GOVREGULAR, WOMENLAB.

Adding Democracy

Adding DEMOCRACY, the solution covers more countries, even if coverage and consistency are still low: Again, I chose a low threshold of 0.8 in order to reach a higher coverage. The sufficient solution for the positive outcome (NDGROWTH) can be found in fsQCA output 12, while fsQCA output 13 displays the solutions for ~NDGROWTH.

Despite these first insights, the model does not yet reflect the sample well enough to allow for conclusions: The coverage is too low. Testing for necessary conditions, the resource conditions were close to being necessary conditions, and there is hope that these conditions help to improve the coverage.

Adding the Resource Conditions

As a nearly necessary condition, the addition of RENT to the model should improve consistency and coverage. This is the case even if the consistency threshold is set at 0.85. The results can be taken from fsQCA outputs 14 and 15.

FsQCA Output 12 Basic Model, DEMOCRACY Tested for NDGROWTH, enlarged sample

NDGROWTH = f(INVESTMENT, POLSTAB, GOVREGULAR, WOMENLAB, DEMOCRACY) *Complex solution*

Frequency cutoff: 1.0; consistency cutoff: 0.836066

	Raw coverage	Unique coverage	Consistency
POLSTAB*GOVREGULAR*WOMENLAB*DEMOCRACY	0.436090	0.436090	0.801105

Solution coverage: 0.436090

Solution consistency: 0.801105

*Cases with greater than 0.5 membership in term polstab*govregular*womenlab*democracy: Norway (0.97,0.6), Netherlands (0.92,0.6), Canada (0.91,0.8), Botswana (0.9,1), USA (0.9,0.7), UK (0.89,0.8)*

FsQCA Output 13 Basic Model, DEMOCRACY Tested for ~NDGROWTH, enlarged sample

~NDGROWTH = f(INVESTMENT, POLSTAB, GOVREGULAR, WOMENLAB, DEMOCRACY) *Intermediate solution*

Frequency cutoff: 1.0; consistency cutoff: 0.823204

	Raw coverage	Unique coverage	Consistency
~DEMOCRACY*~WOMENLAB*~GOVREGULAR*~POLSTAB*INVESTMENT	0.261417	0.150394	0.838384
~DEMOCRACY*WOMENLAB*~GOVREGULAR*POLSTAB*~INVESTMENT	0.117323	0.044094	0.823204
DEMOCRACY*WOMENLAB*~GOVREGULAR*~POLSTAB*INVESTMENT	0.187402	0.075591	0.859206

Solution coverage: 0.384252

Solution consistency: 0.860670

Cases with greater than 0.5 membership in term ~democracy~womenlab*~govregular*~polstab*investment: Iran (0.93,0.7), Saudi-Arabia (0.55,1)*

*Cases with greater than 0.5 membership in term ~democracy*womenlab*~govregular*polstab*~investment:*
Gabon (0.55,0.6)

*Cases with greater than 0.5 membership in term democracy*womenlab*~govregular*~polstab*investment: Venezuela (0.6,0.7), Ecuador (0.53,0.8)*

FsQCA Output 14 Testing Basic Model, DEMOCRACY, and RENT for NDGROWTH, enlarged sample

NDGROWTH = f(RENT, WOMENLAB, DEMOCRACY, GOVREGULAR, POLSTAB, INVESTMENT) *Intermediate solution*

Frequency cutoff: 1.0; consistency cutoff: 0.860140

	Raw coverage	Unique coverage	Consistency
~GOVREGULAR*WOMENLAB*~DEMOCRACY*~INVESTMENT*~RENT	0.189474	0.105263	0.761329
GOVREGULAR*~WOMENLAB*~DEMOCRACY*INVESTMENT*~RENT	0.256391	0.046617	0.919137
GOVREGULAR*WOMENLAB*DEMOCRACY*POLSTAB*~RENT	0.427820	0.044361	0.883540
GOVREGULAR*~DEMOCRACY*POLSTAB*INVESTMENT*~RENT	0.218797	0.000000	0.906542
GOVREGULAR*WOMENLAB*POLSTAB*INVESTMENT*~RENT	0.390977	0.011278	0.943739

Solution coverage: 0.718797

Solution consistency: 0.841549

*Cases with greater than 0.5 membership in term ~govregular*womenlab*~democracy*~investment*~rent: Gabon (0.83,0.4), Nigeria (0.73,0.4), Indonesia (0.69,1)*
Cases with greater than 0.5 membership in term govregular~womenlab*~democracy*investment*~rent: Oman (0.75,0.8), Dubai (0.71,1), Bahrain (0.61,1)*
*Cases with greater than 0.5 membership in term govregular*womenlab*democracy*polstab*~rent: Netherlands (0.92,0.6), Canada (0.91,0.8), Botswana (0.9,1), USA (0.9,0.7), UK (0.89,0.8), Norway (0.61,0.6)*
Cases with greater than 0.5 membership in term govregular~democracy*polstab*investment*~rent: Oman (0.75,0.8), Dubai (0.71,1), Malaysia (0.53,1)*
*Cases with greater than 0.5 membership in term govregular*womenlab*polstab*investment*~rent: Canada (0.91,0.8), Netherlands (0.77,0.6), Botswana (0.7,1), Malaysia (0.68,1), USA (0.65,0.7), Norway (0.61,0.6)*

The test for the positive outcome given in fsQCA output 18 reaches an acceptable consistency of 0.84, while coverage is also not too low. All positive GCC performers are covered. Abu Dhabi, Kuwait, Qatar, and Saudi Arabia as well as Brunei are covered by the test for the negative outcome.

For the negative outcome, consistency is again rather low and the solution terms are long, as output 15 demonstrates. Looking at the parsimonious solution, it seems that WOMENLAB is not an important ingredient of the solutions for the negative outcome. The same analysis without WOMENLAB brings indeed better solutions as output 16 shows.

FsQCA Output 15 Testing Basic model, DEMOCRACY, and RENT for ~NDGROWTH, enlarged sample

~NDGROWTH = f(RENT, DEMOCRACY, WOMENLAB, GOVREGULAR, POLSTAB, INVESTMENT) *Intermediate solution*

Frequency cutoff: 1.0; consistency cutoff: 0.852041

	Raw coverage	Unique coverage	Consistency
RENT*~DEMOCRACY*~WOMENLAB*GOVREGULAR*POLSTAB	0.262205	0.021260	0.840909
RENT*~DEMOCRACY*GOVREGULAR*POLSTAB*INVESTMENT	0.283465	0.042520	0.882353

Solution coverage: 0.304724

Solution consistency: 0.860000

Cases with greater than 0.5 membership in term rent~democracy*~womenlab*govregular*polstab: Qatar (0.86,0.9), Abu Dhabi (0.85,0.7), Kuwait (0.55,1)*
Cases with greater than 0.5 membership in term rent~democracy*govregular*polstab*investment: Brunei (0.88,1), Qatar (0.86,0.9), Kuwait (0.55,1)*

FsQCA Output 16 Testing Basic Model and DEMOCRACY, without WOMENLAB for ~NDGROWTH, enlarged sample

~NDGROWTH = f(GOVREGULAR, RENT, DEMOCRACY, POLSTAB, INVESTMENT) *Intermediate solution*

Frequency cutoff: 1.0; consistency cutoff: 0.859712

	Raw coverage	Unique coverage	Consistency
GOVREGULAR*RENT*~DEMOCRACY*POLSTAB	0.313386	0.288189	0.863340
~GOVREGULAR*~RENT*DEMOCRACY*~POLSTAB*INVESTMENT	0.188189	0.162992	0.859712

Solution coverage: 0.476378

Solution consistency: 0.871758

*Cases with greater than 0.5 membership in term govregular*rent*~democracy*polstab: Brunei (0.93,1), Qatar (0.86,0.9), AbuDhabi (0.85,0.7), Kuwait (0.55,1)*
Cases with greater than 0.5 membership in term ~govregular~rent*democracy*~polstab*investment: Ecuador (0.75,0.8), Venezuela (0.68,0.7)*

RESERVES also came close to a necessary condition that could improve the outcome. Contrary to the analysis of the GGC states only, RENT and RESERVES are no more covariant. Cases such as Norway or Canada especially have RESERVES values higher than 0.5 and RENT fuzzy values lower than 0.5. The results for the basic model including RESERVES now can be taken from fsQCA outputs 17 and 18.

FsQCA Output 17 Testing Basic model, DEMOCRACY, and RESERVES for NDGROWTH, enlarged sample

NDGROWTH = f(RESERVES, WOMENLAB, GOVREGULAR, DEMOCRACY, POLSTAB, INVESTMENT) *Intermediate solution*

Frequency cutoff: 1.0; consistency cutoff: 0.854167

	Raw coverage	Unique coverage	Consistency
~RESERVES*~GOVREGULAR*WOMENLAB*~DEMOCRACY*~INVESTMENT	0.191729	0.105263	0.791925
~RESERVES*GOVREGULAR*WOMENLAB*DEMOCRACY*POLSTAB	0.371429	0.042105	0.914815
GOVREGULAR*WOMENLAB*DEMOCRACY*POLSTAB*INVESTMENT	0.383459	0.062406	0.836066
~RESERVES*GOVREGULAR*~WOMENLAB*~DEMOCRACY*~POLSTAB*INVESTMENT	0.166165	0.078195	0.876984

Solution coverage: 0.633834

Solution consistency: 0.802093

Cases with greater than 0.5 membership in term ~reserves~govregular*womenlab*~democracy*~investment:*
Nigeria (0.73,0.4), Indonesia (0.69,1), Gabon (0.67,0.4)
*Cases with greater than 0.5 membership in term ~reserves*govregular*womenlab*democracy*polstab:*
Botswana (0.9,1), UK (0.89,0.8), Netherlands (0.87,0.6), USA (0.67,0.7)
*Cases with greater than 0.5 membership in term govregular*womenlab*democracy*polstab*investment:*
Norway (0.97,0.6), Canada (0.91,0.8), Netherlands (0.77,0.6), Botswana (0.7,1), USA (0.65,0.7)
*Cases with greater than 0.5 membership in term ~reserves*govregular*~womenlab*~democracy*~polstab*investment: Bahrain (0.77,1)*

FsQCA Output 18 Testing Basic Model, DEMOCRACY, and RESERVES for ~NDGROWTH, enlarged sample

~NDGROWTH = f(RESERVES, DEMOCRACY, WOMENLAB, GOVREGULAR, POLSTAB, INVESTMENT)		*Intermediate solution*	
Frequency cutoff: 1.0; consistency cutoff: 0.813397			
	Raw coverage	**Unique coverage**	**Consistency**
~DEMOCRACY*~WOMENLAB*~GOVREGULAR* ~POLSTAB*INVESTMENT	0.261417	0.152756	0.838384
DEMOCRACY*WOMENLAB*~GOVREGULAR* ~POLSTAB*INVESTMENT	0.187402	0.078740	0.859206
Solution coverage: 0.340157			
Solution consistency: 0.855446			

Cases with greater than 0.5 membership in term ~democracy~womenlab*~govregular*~polstab* investment: Iran (0.93,0.7), Saudi-Arabia (0.55,1)*
*Cases with greater than 0.5 membership in term democracy*womenlab*~govregular*~polstab* investment:*
Venezuela (0.6,0.7), Ecuador (0.53,0.8)

Testing for negative diversification results (~NDGROWTH), RESERVES does not seem to have any explanatory power. It appears neither in the complex/intermediate solution (identical) nor in the parsimonious solution. The coverage of the solution is considerably lower than with RENT: in fact, only four cases are covered (compared to eight with RENT). Partly this could have been expected due to the lower consistency for necessity.

Adding Natural Resource Exports in GNP (SXP)

In the preceding analysis, the inclusion SXP and natural resource rents as a percentage of government income (RENTIERI) did not have the same results as RENT and RESERVES. This was explained by the fact that they are resource dependence measures. Nevertheless, a test of these conditions may be useful to confirm this suspicion. The results can be taken from fsQCA outputs 19 and 20.

With the inclusion of countries like Nigeria, for example, new types of countries have been added to the sample. Nigeria is not extremely resource rich, but has attained a 0.87 on the rentier scale.

Hence, these two conditions are more likely to lead to contradictions.

FsQCA Output 19 Testing Basic Model, DEMOCRACY, and SXP for NDGROWTH, enlarged sample

NDGROWTH = f(SXP, GOVREGULAR, DEMOCRACY, WOMENLAB, POLSTAB, INVESTMENT) *Intermediate solution*

Frequency cutoff: 1.0; consistency cutoff: 0.860781

	Raw coverage	Unique coverage	Consistency
~SXP*GOVREGULAR*DEMOCRACY*WOMENLAB*POLSTAB	0.433835	0.052632	0.820768
~SXP*GOVREGULAR*WOMENLAB*POLSTAB*INVESTMENT	0.396993	0.015789	0.864157

Solution coverage: 0.449624

Solution consistency: 0.824828

*Cases with greater than 0.5 membership in term ~sxp*govregular*democracy*womenlab*polstab: Norway (0.97,0.6),Netherlands (0.92,0.6), Canada (0.91,0.8), Botswana (0.9,1), USA (0.9,0.7), UK (0.89,0.8)*
*Cases with greater than 0.5 membership in term ~sxp*govregular*womenlab*polstab*investment: Norway (0.98,0.6), Canada (0.91,0.8), Netherlands (0.77,0.6), Botswana (0.7,1), Malaysia (0.68,1), USA (0.65,0.7)*

FsQCA Output 20 Testing Basic Model, DEMOCRACY, and SXP for ~NDGROWTH, enlarged sample

~NDGROWTH = f(SXP, DEMOCRACY, WOMENLAB, GOVREGULAR, POLSTAB, INVESTMENT) *Intermediate solution*

Frequency cutoff: 1.0; consistency cutoff: 0.848837

	Raw coverage	Unique coverage	Consistency
SXP*~DEMOCRACY*~WOMENLAB*~GOVREGULAR*~POLSTAB*INVESTMENT	0.113386	0.077165	0.917197
~SXP*DEMOCRACY*WOMENLAB*~GOVREGULAR*~POLSTAB*INVESTMENT	0.172441	0.136220	0.848837

Solution coverage: 0.249606

Solution consistency: 0.890449

Cases with greater than 0.5 membership in term sxp~democracy*~womenlab*~govregular*~polstab*investment: Saudi Arabia (0.55,1)*
*Cases with greater than 0.5 membership in term ~sxp*democracy*womenlab*~govregular*~polstab*investment:*
Venezuela (0.6,0.7), Ecuador (0.53,0.8)

Natural resource export in GNP (SXP) is certainly not a useful condition for the improvement of the model. While it is part of the solutions for the negative outcome, it leads to a low coverage and is part of the solution as both present and not present. It seems that the fact that SXP reflects different situations is even more obvious in this sample.

Adding Rent as a Percentage of Government Income (RENTIERI)

The same holds true for RENTIERI. Summarized results can be taken from tables 55 and 56 below. Coverage decreases and for the negative outcome, only two countries are covered. While RENTIERI is part of the solution, it is more likely that other factors are more important.

Interpretation

Again, interpreting these various solutions and models calls for a summary of the results. Given the increased number of cases and conditions, the solutions are more complex. This is why I will use the parsimonious solutions for a short comparison. As explained above, parsimonious solutions rely on assumptions about the logical remainders. They will not be the sole basis for the conclusions, but they display the most important conditions and models. Hence, they are a good starting point to interpret the results.

Firstly, one condition turned out to be nearly necessary: ~RENT ← NDGROWTH

The consistency was at 0.91 and rather low. As such, it should not be excluded from the test for sufficient conditions. Plainly spoken, this means that countries with an extremely high annual average resource rent do not fulfill this necessary condition for NDGROWTH and should have failed to diversify successfully (~NDGROWTH).

The results of the tests for sufficient conditions can be summarized in the following tables 53 and 54.

Before interpreting these solutions it should be noted that six countries are already explained by a model without any resource conditions: These are the Western democracies of The Netherlands, Norway, the UK, USA, and Canada. Western democracies stand out among the 25 countries (or 26 cases) because they are explained without using any resource condition. The basic model, including democracy, offers the solution POLSTAB*DEMOCRACY. Only one other country is explained by this sufficient solution: Botswana, a notorious special case, and a prime example for those who claim that development and democracy in Africa are possible (at least in a state with only two million inhabitants).

In these Western democracies, the combination of political stability and democracy seems to be conducive to growth. At times, members of parliament may use their power to block reforms and politicians may resort to pre-election goodies, but the impact of these factors does not annihilate the meaningful role of the parliament in shaping and legitimizing government decisions.

Table 53 Sufficient Solutions for the Basic Model, DEMOCRACY and Resource Conditions, Tested for NDGROWTH (enlarged sample, parsimonious solutions)

	SOLUTIONS FOR → NDGROWTH	Countries/Coverage
Basic Model + DEMOCRACY	POLSTAB*DEMOCRACY	NL, NOR, CND, BOT, USA, UK
	Solution consistency: 0.80	Coverage: 0.46
Adding RENT	~RENT*POLSTAB	NL, BOT, CND, USA, UK, MAL, OM, DUB, NOR, GAB
	~RENT*~DEMOCRACY~GOVREGULAR	OM, DUB, BAH, MAL
	~DEMOCRACY*WOMEBLAB*~GOVREGULAR	GAB, NGA,IDS
	~INVESTMENT*WOMENLAB*~GOVREGULAR	GAB, NGA, IDS
	~INVESTMENT*~DEMOCRACY*WOMENLAB	GAB, NGA, IDS
	~RENT*~DEMOCRACY*WOMENLAB	GAB, NGA, IDS, MAL
	solution consistency: 0.84	solution coverage: 0.72
Adding RESERVES	POLSTAB*DEMOCRACY	NL, NOR, CND, BOT, USA, UK
	~DEMOCRACY*WOMENLAB*~RESERVES	NGA, IDS, GAB, MAL
	~POLSTAB*~DEMOCRACY*GOVREGULAR	BAH
	solution consistency: 0.77	solution coverage: 0.68
Adding SXP	POLSTAB*WOMENLAB*GOVREGULAR*~SXP	NL, BOT, NOR, CND, USA, UK, MAL
	solution consistency: 0.83	solution coverage: 0.45
Adding RENTIER	POLSTAB*~RENTIERI	NOR, NL, CND, USA, UK, MAL, DUB, GAB
	~DEMOCRACY*WOMENLAB*~RENTIERI	IDS, GAB, MAL
	~INVESTMENT*~GOVREGULAR*WOMENLAB*~RENTIERI	IDS, GAB
	solution consistency: 0.83	solution coverage: 0.6

A list of abbreviations can be found in Appendix A.

Table 54 Sufficient Solutions for the Basic Model, DEMOCRACY and Resource Conditions, Tested for ~NDGROWTH (enlarged sample, parsimonious solutions)

	SOLUTIONS FOR → ~NDGROWTH	Countries/coverage
Basic model + DEMOCRACY	INVESTMENT*~GOVREGULAR	IRN, ECU, SA, VEN
	POLSTAB*~GOVREGULAR	GAB
	~INVESTMENT*POLSTAB*WOMENLAB*~DEMOCRACY	GAB
	solution consistency: 0.83	Coverage: 0.45
Adding RENT	RENT	ADH, QAT, BRU, KWT
	solution consistency: 0.81	solution coverage: 0.42
Adding RESERVES	INVESTMENT*~WOMENLAB*~GOVREGULAR	IRN, SA
	INVESTMENT*~DEMOCRACY*~GOVREGULAR	IRN, SA
	solution consistency: 0.83	solution coverage: 0.33
Adding SXP	~GOVREGULAR*DEMOCRACY	VEN, ECU
	~GOVREGULAR*SXP	SA
	~POLSTAB*SXP	SA
	INVESTMENT*~GOVREGULAR*WOMENLAB	VEN, ECU
	INVESTMENT*~POLSTAB*WOMENLAb	VEN, ECU
	solution consistency: 0.83	solution coverage: 0.38
Adding RENTIERI	INVESTMENT*~WOMENLAB*~GOVREGULAR	IRN, SA
	INVESTMENT*~DEMOCRACY*~GOVREGULAR	SA, IRN
	solution consistency: 0.83	solution coverage: 0.33

A list of abbreviations can be found in Appendix A.

For both the negative and the positive outcome, adding the annual average resource rent *per capita* (RENT) to the model clearly brought the best solutions. Due to their importance, these solutions have been summarized again in table 55.

Table 55 Basic Model, DEMOCRACY and ~RENT (sufficient solutions, parsimonious)

~RENT*POLSTAB → NDGROWTH

~RENT*~DEMOCRACY~GOVREGULAR → NDGROWTH

~DEMOCRACY*WOMEBLAB*~GOVREGULAR → NDGROWTH

~INVESTMENT*WOMENLAB*~GOVREGULAR → NDGROWTH

~INVESTMENT*~DEMOCRACY*WOMENLAB → NDGROWTH

~RENT*~DEMOCRACY*WOMENLAB → NDGROWTH

RENT → ~NDGROWTH

Furthermore, these countries are developed countries. The government does not plan development from scratch: It tries to correct development, give incentives, decide over welfare and education, because the way of development had been set much earlier. The argument that government needs a strong, independent hand to implement large-scale development plans does not hold in this case.

Political stability is not "bought" by governments with vast privileges for the citizens, but is the result of the fact that a majority of the populations identifies with the political system. If not, there are peaceful channels for critique such as a free press, free NGOs, associations, and demonstrations. Radical opposition is present, but still marginal. Meritocratic principles are not entirely deactivated by welfare. Even if a social net of benefits exists, at least in the European states, this is not abused by a majority of the population. In short, as imperfect as Western democracy may be, it functions quite well when compared to other regions of the world.

These five states offer a wide bandwidth of annual average resource rents: The US, Canada, the Netherlands and the UK range from US $300 *per capita* (as of 2000) to roughly US $760 US (in Canada), while Norway receives more than US $3500, which is roughly the category of Bahrain but still far from Qatar's US $10,600. None, not even Norway, is a rentier state. When oil and gas windfalls occurred, these states were already developed with a developed tax system. Moreover, the windfall gains were not high enough to abolish taxation. Nonetheless, given that there were well-functioning manufacturing sectors in these liberal economies, Dutch disease cannot be excluded.

In the case of the Netherlands, Dutch disease was very short-lived, and after the first years of the resource windfall, the manufacturing sector even grew again. In the case of Norway, however, it seems that manufacturing suffered from resource windfalls and failed to recover: It still plays a relatively small role in the Norwegian economy (2004: 19 percent of exports compared to 70 percent in the Netherlands). My overall calculation of non-oil growth does not make a difference between the growth of the manufacturing sector or the service sector, but Norway's non-oil growth of 1.3 percent (fuzzy value 0.6) is comparatively low, and there are weak signs of an ongoing Dutch disease effect. Nonetheless, it seems that Norway was able to avoid most other negative effects of natural resource abundance: Rent-seeking is not much of a problem. The percentage of students enrolled in higher education is very high and has increased rapidly in the last decades. Oil rents *are* used to cover budget deficits although the budget is still not inflated: Government and bureaucracy are rather small and efficient. There is a very well-filled pension fund, which is supposed to help finance the pensions of elderly people that can no longer be paid by the younger generation. Thus Norway can afford the luxury of leveling-out problems of an aging population by resource rents. Furthermore, this fund is today managed by the Bank of Norway and cannot be used arbitrarily by politicians. Gylfason concludes that "Norway's tradition of democracy since long before the advent of oil has probably helped immunize the country from the ailments that inflict most other oil-rich nations." (Gylfason 2006: 4)

From this example it becomes clear that the stage of development of a country that experiences resource windfalls is an important factor. While previous development and a liberal economy make Dutch disease possible, all other "diseases", mainly resulting from ill-conceived incentives, can be avoided by democratic institutions. Despite Dutch disease, Norway has a viable economy.

Another condition playing a prominent part in sufficient solutions for these developed states is GOVREGULAR. A good regulatory quality and good governance (and also a rule of law and control of corruption) often go hand in hand with democracy and political stability. This is the relationship as assumed by new institutionalist economists: Democracies guarantee viable and sound institutional environments. From other examples it becomes clear that the opposite is not necessarily true: A viable institutional environment does not presuppose a democracy – but probably, the costs for sound institutions are higher in autocratic states.

When adding the resource conditions, all developed countries have low values (~RENT, ~RESERVES, ~SXP and ~RENTIER). Nevertheless, these factors are not necessary to explain the reasons why they had a sound non-oil growth. Here, political stability and democracy (POLSTAB and DEMOCRACY) were enough to explain the outcome. In such an environment, natural resource abundance obviously does not play a crucial role.

As a last point, I should mention that the fuzzy values for the outcome of these Western countries were mostly below 1, but above 0.6. Economic growth of the non-resource sectors, therefore, was not perfect either.

Another five countries are explained without any use of a resource condition: Saudi Arabia, Iran, Venezuela, Ecuador, and Gabon.

During the entire analysis, Saudi Arabia and Iran were covered by the same solutions. Iran, Ecuador, Saudi Arabia, and Venezuela made a lot of investments (calculated as share of normal GDP), but had low regulatory quality and also low political stability. This combination of factors correlates with a negative outcome in all countries, even if Iran and Saudi Arabia are not democratic and have a low female labor force participation rate (~WOMENLAB), while Ecuador and Venezuela are at least partly democratic and have rather high female labor force participation rates (WOMENLAB). It seems that these last two factors do not matter anymore once regulatory quality and political stability are bad. The same holds true of investments, of which the positive effects seem to equally depend on a good regulatory quality. This result confirms the results of the previous analysis when governance and regulatory quality (GOVREGULAR) formed a necessary condition.

As for Gabon, bad regulatory quality and bad investments correlate with a low growth performance despite political stability.

To sum up, among the conditions INVESTMENT, POLSTAB and GOVREGULAR, it is not enough to fulfill only one. To speak in terms of QCA, it is sufficient for a bad diversification when regulatory quality/governance or investments are bad, as the parsimonious solutions suggests. WOMENLAB and DEMOCRACY seem to be rather marginal issues and can no longer change the outcome once political stability, regulatory quality, or investments are low.

Once the model includes the annual average resource rent *per capita* (RENT), this condition becomes an important factor. The absence of high rents is part of all sufficient solutions (intermediate and complex) needed for a negative diversification result (~NDGROWTH). The absence of high rents combined with political stability is one possible path to a good diversification result (NDGROWTH). This concerns the Western democracies and Botswana, but also Oman, Malaysia and Dubai, which are all institutionally sound and politically stable although not democratic, with Malaysia coming closest to a democratic country with a fuzzy value of 0.47.

Several times, the absence of democracy (~DEMOCRACY) is part of a sufficient solution for successful economic growth (NDGROWTH). Again, democracy shows its ambivalent role: If combined with political stability, it is condusive to growth. If democracy is bought at the price of political instability it may be better to be outright autocratic than "nearly" democratic. Indeed, apart from the Western democracies and Botswana, all other sufficient solutions for the positive outcome (NDGROWTH) include the absence of democracy (~DEMOCRACY) and not DEMOCRACY. Of course, a third case – namely autocratic unstable states – is equally possible, as in the case of Iraq for example.

In autocratic rentier states, the costs of political stability may be even worse for the economy than would be limited political instability. It seems that certain kinds of political instability do not harm the economy as much as inefficiency, due to costly welfare systems set in place to calm the population. However, this interpretation cannot solely rely on these results. While the results assert that political stability is sometimes and sometimes not part of a sufficient solution, the conditions fail to reflect the underlying reasons for political situation.

The importance of regulatory quality and governance (GOVREGULAR) is also confirmed. While the Western democracies have all received high ratings for these indicators,

GOVREGULAR countries are not part of the solution because they are all politically stable and democratic. Good regulatory quality seems to be a natural correlate of these characteristics. In those countries that are not democratic and unstable, however, GOVREGULAR becomes the crucial factor. In the absence of democracy, a low average annual resource rent and a sound institutional environment (GOVREGULAR, and also RULEOFLAW) are sufficient for NDGROWTH.

The test for sufficient conditions for NDGROWTH also includes one solution with ~GOVREGULAR.

The term

~GOVREGULAR*WOMENLAB*~DEMOCRACY*~INVESTMENT*~RENT → NDGROWTH

concerns the countries Gabon, Nigeria, and Indonesia. The parsimonious solution reduces this term to various combinations of three conditions, while WOMENLAB is the common denominator. Interestingly, Gabon and Nigeria are actually negative performers, even if they have fuzzy values of 0.4, which is close to a positive outcome. Thinking in terms of a xy-plot and the diagonal, sufficient conditions and outcomes may be too far away from the diagonal despite 0.4 being actually below. However, the distance is reflected in a low consistency of this solution (0.76). Nonetheless, it seems that in Africa, female labor force participation (WOMENLAB) plays a special and crucial role in preventing a disastrous growth performance, which would have to be expected due to low regulatory quality and low investments. This also holds true for Indonesia, which was – compared to Malaysia – institutionally weak and less stable, at least during the earlier part of the time period under dictatorship. The fact that both Indonesia and Malaysia reached such a high growth rate, however, is not sufficiently explained by these models. As the cases of Nigeria and Gabon indicate, it is rather contradictory that Indonesia should have reached such a high growth rate. This apparent contradiction is evocative of the debate about "Asian values" - a supposed high work ethic among Asians which is thought to contrast with a rather low work ethic in Africa or the Middle East. Asian values have been used to justify dictatorship and a lack of civil liberties and political rights, while putting priority on work for the community and economic well-being. This is reminiscent of the high growth rates of the Chinese economy which have continued despite the highly autocratic nature of its political system. In Malaysia, the Chinese minority plays a disproportionately large role in the economy despite privileges granted exclusively to the Malay majority. (Stauth 2002)

Turning to solutions for the less successful diversifiers (~NDGROWTH), the parsimonious solution is very short: high annual *per capita* resource rents (RENT). The coverage of this parsimonious solution is at 0.42 and rather low. Interestingly, it covers the states from the previous analysis: Abu Dhabi, Qatar, and Kuwait and also Brunei, which is very similar to the GCC oil monarchies: It is a very small, autocratic monarchy, possessed of massive fossil fuel reserves, and with all the gratifications for the nationals known to the GCC states.

It is also worth looking closer at the importance of RENT as a condition. Contrary to the previous analysis, RENT turned out to be more important than RESERVES. Several cases have high reserves but low rents. This deviation stems from different factors. In the case of Canada, for example, production costs of oil are high, lowering the amount of rents, while reserves appear to be enormous. If oil prices are low, oil production may not be lucrative at all. In Norway, oil and gas production is "husbanded" by discipline and production costs are also higher than in the Middle East. (Etherington 2010) In short: Governments influence their natural resource production and hence their rentier income while they cannot influence the amount of their reserves. The fact that rents are comparatively low already reflects a prudent approach of governments that are not dependent on this income and not dependent on other governments persuading them to produce more than they deem healthy for their economy.

It is also noteworthy that the amount of reserves alone does not have a direct effect on the economy or the welfare system. Only if incentives are indeed lowered by the amount of reserves available can reserves have effects on the economy. As demonstrated above, Western governments have mostly avoided this "incentives trap", while the GCC states apparently have not. The fact that annual average resource rents are more important than the reserves in this sample can be explained by contrasting the GCC with Western developed countries.

Methodological Interpretation

The enlargement of the sample had an ambivalent effect on the results. While some of the previous results were hardened and specified, others were diluted. The latter is true for the necessary conditions: The importance of regulatory quality was confirmed, but only as a part of a sufficient solution and not as necessary solutions anymore.

Hence, the larger sample clearly corroborates that the effects of conditions on the outcome depends upon the context. Democracy and political stability turned out to be highly dependent on further circumstances. Depending on the context, political stability and democracy correlated with a good or a bad diversification result. This finding corresponds to problematic behavior of variables on political stability and democracy in regression analyses. It also became clear that the solutions for the Western developed nations did not need any resource condition for an explanation. Similarly, other sufficient solutions did not rely on reserves or the average annual resource rent *per capita*. Thus, there are countries disposing of a comparable amount of natural resources where natural resources matter, and there are others where they do not matter. Such a result could not have been produced by a regression analysis: Here, rents matter or they do not matter. At best, previous regression analyses have shown that the significance of natural resource variables depended not only on the operationalization, but also on the sample size. Nevertheless, path dependence and equifinality also revealed disadvantages: Solution terms were sometimes too long to put these interactions into short summaries. The complexity of solutions also made interpretation more difficult.

The amount of cases and conditions also had another effect: The model can be tested, but it becomes increasingly difficult to see how different conditions could change the coverage. In the first analysis, it was relatively easy to see why certain cases were not covered. In the larger sample, it is no longer possible for the researcher to retrace the relationship between data, cases, and solutions. The analysis becomes less case centered the more cases and conditions involved.

It is also noteworthy that the coverage of the solutions reflects the theoretical background: The theories which marked my choice of variables and conditions were, on the one hand, growth theories. Their assumptions came from insights about Western economies. On the other hand, I tried to enlarge the theoretical background by adding theories about rentier states. The Western democracies and the GCC states are the states which are best covered by the analysis. This is less true for the African states and South America, and one group is especially and persistently absent: non-GCC Arab countries such as Algeria, Iraq, and Egypt. According to my calibration, they all have rather low RENT and RESERVES values but have different outcomes: Two are relatively good performers (Algeria, Egypt) and one is a disastrous performer (Iraq). They all have very low values for political stability and also institutional quality, and again – looking at the background of political instability – it becomes clear that the scale and reasons for political instability in Egypt and Algeria can hardly be compared to the situation in Iraq. Furthermore, they are all autocratic regimes. Their democracy scores resemble those of the GCC states, but their type of autocracy is different. All three countries experienced revolutions, and have abolished their monarchies or have never had one. This different nature of authoritarianism seems to be intrinsically linked to the lack of political stability. Again, the conditions DEMOCRACY and POLSTAB give rise to doubts about the measurability of such heterogeneous situations and the use of qualitative rather than quantitative regime classifications could be useful. In this context, Hadenius and Teorell (Hadenius/Teorell 2007: 143-156)[5] have recently proposed a new classification of authoritarian regime types which further develops the work of Barbara Geddes (1999).

It is possible that a theoretical focus on these states would have drawn attention to different theories and different political, social, and historic issues. This may have led to a better coverage, but also to more conditions. With a sample of 25 (26) countries, the number of conditions is already at a limit. The complex and intermediate solutions are often too long to be interpreted with simple statements. Hence, the sample may be too diverse despite the common focus on oil rents. More homogenous samples such as the countries of the European Union or the countries of the Arab League may yield a better coverage despite a smaller number of conditions. This was also an insight from the operationalization of conditions: Several conditions were applied to the entire sample against better judgment.

6

Conclusion

It had been the primary aim of this study to investigate whether the GCC states have made any progress in developing their non-oil economies and to what extent the individual countries were successful or not. The analyses furthermore investigated the factors, and as a background, the theories that might explain the failure or success of single countries. In doing so, my analysis found that special attention was paid to the role of natural resource dependence and abundance for this process. Furthermore, one major idea behind this study was to also open up new methodological perspectives for the analysis of groups of economies with special distinctive features.

With regard to the first question, the GCC countries show heterogeneous success stories when it comes to economic diversification. They have all made some progress, even if none of them reached independence from its oil- and gas sectors. It is, to be sure, unclear how such independence should look at a time when oil and gas are still available, produced and sold at a relatively high price.

The more successful "diversifiers" are Bahrain, Dubai, and Oman. The rather "modest" diversifiers are Qatar and Abu Dhabi, while Saudi Arabia and especially Kuwait obviously faced huge difficulties during the period under scrutiny. These are the results of the admittedly very simplified calculations of average non-oil growth, which nevertheless conform to expert assessments in the qualitative literature, and results obtained from sectoral data. (al-Kuwari 2013) Given the huge data problems this was not a matter of course, and my data on non-oil growth rates as well as common data on resource dependence (natural resource rents as a percentage of GDP or total government income) have to be interpreted with caution. The rentier multiplier effect is probably very strong and not included in such calculations.

The factors which seemed to be important in the case studies turned out to be decisive in the comparative fsQCA analyses, although not always in the regressions. The regressions only resulted in the importance of investments, initial GDP, fertility rates, and institutions. Investments were also important in the fsQCA analyses. While the variable had to be changed into "good investment behavior" (INVESTCAPITA), it was very close to being both a necessary and sufficient solution, and as such this should be kept in mind. Interestingly, only Bahrain, Dubai and Oman decided to invest most of their rents locally.

For Bahrain, Dubai and Oman, incentives to diversify have always been very high due to comparatively modest rents and reserves. Dubai and Bahrain diversified early and could avoid an extreme dependence on their natural resources, even if they remained dependent on oil and gas. Oman, as a late diversifier, was at the outset dependent on its oil and gas because of the lack of a viable economy. However, low reserves were incentive enough to seriously tackle its resource dependence. All three states show good regulatory quality, governance, and rule of law, while political stability seems to matter little in such an institutional setting. None of them has a meaningful parliament and the development process remains firmly in the hands of the "royal" families.

While the three successful diversifiers have all followed the same path, the less successful diversifiers are more heterogeneous.

Abu Dhabi, Qatar, and Kuwait have less incentive to diversify rapidly due to their extremely high rents and ample reserves. They did not invest most of their rents locally. Although they have a good institutional setting and a strong regulatory quality, they have been very careful or even hesitant in choosing their path to development. The observance of Islamic values, the avoidance of social disruptive policies, and more prudent investment in new sectors are the result. In Kuwait, the will to diversify is also obstructed by the parliament, which displays a strong interest in the continuity of habitual privileges and rent seeking. Despite several dissolutions and new elections, the role of the parliament is unlikely to change in the near future. (Al-Zumai 2013)

In Saudi Arabia, incentives to diversify are strong, but regulatory quality and governance are rated low, while rule of law is only rated as "modest". The combination of low rents and an institutional setting is perceived as weak, thus seems to lead equally to a negative diversification result. Apart from the size of the country, the strong role of conservative Islam – or at least the perception of it – may play a role in the bad institutional setting and low regulatory quality. The fsQCA analysis showed a correlation of the two aspects. In the larger sample, the paths to development of Iran and Saudi Arabia were suspiciously similar although no condition on Islam was included. The gap between an ideological ideal and the basic necessities of a viable economy may lead to an inconsistent institutional setting where public discourse and reality often fail to coincide. This uncertainty may discourage investors unfamiliar with how business works there.

In general, the fsQCA-analysis of the smaller sample confirmed the case studies. The institutional conditions and the amount of rents and reserves were crucial, including from a comparative point of view. Some aspects highlighted by the case studies, however, could not be operationalized or could not be compared due to limited variance. The fsQCA analysis, even of the small sample, cannot therefore replace proper case studies: It has to draw on them and add explanatory power that the case studies would probably overlook.

In the larger sample, the GCC states were all covered by the resulting solutions, but they formed their own paths to development. This confirms the view that their specificities make them a sample on their own, especially when it comes to such contentious factors like democracy. Other factors, such as investments, Arab society, or the percentage of Muslim inhabitants were less reliable because they were operationalized differently.

The regression, however, only partly reflects the reality of the GCC states. The annual *per capita* resource rent (RENT) and the limited importance of democratic organs (DEMOCRACY) are undeniably important factors in the GCC states, but the regression did not show any significance. On the other hand, political stability (POLSTAB) was a significant variable, but the fsQCA analysis showed that its relationship with growth is not unambiguous. The variables on investments (LOGINVEST) and on initial GDP (LOGNDINITIAL) cannot be fully compared from analysis to analysis. The high significance of the institutional variable, however, was in agreement with both fsQCA analyses.

Despite the plausibility of these results, it should be noted that this comparative study required a fixed time frame. This time frame ends in 2006 for reasons of data availability at the beginning of this study. In the 21st century, however, the diversification strategies of the six GCC states seem to especially copy the Dubai Model: Tourism, international financial centers, attraction of foreign direct investment, and extraordinary cultural attractions. Apparently, Dubai has triggered a regional competition and no other GCC state wants to leave the benefits of diversification to Dubai only. In Qatar, for instance, the non-oil growth rates in the 21st century were considerably higher than before, and a few more years of these growth rates could have changed the overall average non-oil growth rate significantly. (Fasano/Iqbal 2003: Fig. 2)[1] The same is true for Saudi Arabia under King Abdullah. Abdullah dared to take several important steps towards an improvement of the legal environment, the financial sector, and towards more liberty in restricted areas. On the other hand, Dubai – the supposed role model of the region – was hit hard by the international financial crisis in 2008, although it is rather unlikely that a few "bad" years will lead to a negative reevaluation of the entire diversification process. Nonetheless, in the long run, the slow diversifiers could turn out to be the winners in this competition for economic diversification because it remains unclear what the GCC states are preparing for. It might be a future with or without oil. A potential comparative advantage in solar energy might replace the comparative advantage in conventional energy production with fossil fuels. Wars and military expenditures may cause economic disasters, and environmental damage due to industrialization may be have to be taken more seriously than today. In the history of economics, more often than not predictions about future developments have turned out to be wrong, and it is not unthinkable that diversification as it happens today will later appear as unnecessary or prematurely initiated. Although there seems to be widespread agreement that "diversification" is something good and useful, the GCC states may just as well use their comparative adavantage, sell off their oil and gas first, and diversify later.

These comparative analyses are a snap-shot – though a rather long one – based on the conventional wisdom of today. In ten years evaluations may change, not only in their standard but also numerically. Changed outcomes, in turn, could also lead to a different assessment of the importance of variables and conditions. The lack of incentives due to high resource rents could, for instance, be overcome by the regional "competition" for international importance, tourism, and FDI attraction, as well as the fear of leaving the entire potential to the neighbors. This competition has already taken place and has led more to a duplication of projects than to a regional division of labor and a wise, transnational use of linkages. Although the region is heterogenous enough to allow for intra-regional comparative advantages, this potential seems to have been largely neglected so far.

From the methodological point of view, there are further interesting insights. There is a widely used distinction between variable-oriented and case-oriented research approaches. While the former can put more focus on the variables, the latter center on individual cases.

I only partly found evidence for this point of view: From the cases studies to the fsQCA analysis of the small sample, several tradeoffs had to be made concerning operationalization. The mere use of variables or conditions meant a simplification of complex processes. In the larger sample, less attention was paid to the cases, but it would be wrong to assume that, in turn, there was a more thorough handling of the variables that would have improved their quality. On the contrary: Several operationalizations had to be used despite better information, because one single operationalization for all different cases was necessary. For the GCC states, this meant that several results from the first fsQCA analysis could not be reproduced in the second. Furthermore, consistency levels sank and coverage was also lower. Contrary to the regression, however, the fsQCA analysis showed clearly which cases were not reflected in the solution terms.

Therefore, it seems that case-orientation and variable-orientation go hand in hand. In bigger samples, the variables also become less accurate, although more thought may be given to their theoretical underpinning. Variables and cases are inseparably linked, putting the universality of variables and the underlying theories into question.

Furthermore, one of the advantages of methodological triangulation or mixed methods research in general could be a learning process from case studies and small samples to large n-samples. The insight that there is actually no reason to treat the UAE as one case could easily be applied to a regression, and there may be more countries, most likely other federations, where economic policy is decentralized or where economic development is very heterogeneous. Canada, Belgium, and today also Iraq, come to mind in this regard. Of course, splitting these cases requires data on a sub-national level.

The results of the multimethodology in this study also shed new light on the growth and development theories discussed at the beginning. Partly, their limited applicability to the GCC states was foreseeable. I had to acknowledge that there is no proper operationalization to test certain theories and some aspects had to be dropped for the sake of parsimony.

In sum, however, the results from my analyses could add something to these theories, specify them, or tackle problems of operationalization as follows:

Firstly, "old" growth theory was confirmed. Investments were robust in the regression and almost necessary as part of the sufficient solution in the first fsQCA analysis. However, the operationalization of investments turned out to be problematic and the tradeoff of the second fsQCA analyses (the enlarged sample) may explain why the condition INVESTMENT was not as important as in the smaller sample. The same was true for initial GDP. GDP data and investment data is often considered "hard data" as compared to rather "soft" data, based on perceptions and assessments such as the institutional ratings. In this study, the weaknesses of "hard data" were obvious.

Secondly, given the circumstances of the region, human capital could not be confirmed by the empirical comparative analyses. The case studies point to a complex interplay of factors affecting the beneficial aspects of human capital, which was displayed neither by a regression nor by fsQCA. First, not only can labor be imported, but also the technologies do not need to be developed and instead can be bought and applied by foreign experts. The consistent look at what to buy from the West (and increasingly, the East) may in the end forestall genuine local innovation and resulting spillover effects. Second, rentier mentality and low female labor force participation "wastes" investments in education, and precludes positive spillovers and learning-by-doing effects. The assumed positive effects of human capital on endogenous growth are thus much more complex. Probably a stronger focus on this particular issue, including conditions on labor migration, rentier mentality or work motivation, hidden unemployment, and maybe private sector employment of the nationals also, could have provided better results. It should also be noted that high growth rates are often achieved in countries where human capital has just started to grow. Export-led growth, as in China for example, needs a few experts but in the main masses of cheap workers. To sum up, the link between human capital and growth is severely distorted in the GCC states.

As for endogenous development theories, the comparative advantage was a big issue in diversification planning. This "theory" was not tested by a variable but the case studies showed that this comparative advantage suffered from several flaws. The advantage of cheap energy for heavy industries or petrochemicals is undeniable when physical capital is the main factor. However, the lack of natural gas in some countries, and in particular the lack of expertise that has to be imported together with the machinery, decreases this advantage.

Big Push theories are of special importance. In the GCC states, their importance was demonstrated by the qualitative studies as well as by the fact that the variables/conditions on governance and regulatory quality (GOVREGULAR) were crucial. The clear will to improve investment climates and governmental investments into infrastructure have indeed attracted many foreign investors. Nevertheless, the problems of balanced growth became obvious: There were several huge projects that turned out to be white elephants. By adopting short planning intervals (five-year plans) such white elephants could be reduced. The initial euphoria of the Big Push's, however, have often ceded to indecision and mistrust that is costly and inefficient, too. And finally, the calculation did not add up as planned. There was not only an initial Big Push, but governments continue to "push" the industrialization and the development of a viable private sector. Currently, the governments run several programs to lure nationals into the small and medium business segments but face the people's preference for security. (Kolb 2015)

Hence, one may ask whether the governments have provided too much security. Potential private investors do not just have the security that the government provides to the infrastructure and leading-sector investment industries, they also expect fast profits without risk as well as protection from foreign competition. While the GCC states are among the few countries in the world that can afford such a Big Push, they may have overshot the mark. This also concerns subsidies and protective tariffs as propagated by dependency theory, which have come to be regarded as a privilege of the entrepreneur.

Unlike common development theories, operationalization was easier for new economic institutionalism. The important factors of institutional theories were present in the comparative analyses. The variables/conditions of interest were the absence of corruption and rule of law (RULEOFLAW), governance and regulatory quality (GOVREGULAR), political stability (POLSTAB) or the aggregate variable INSTITUTIONS. Democracy (DEMOCRACY/DEMOCRACY) and PARLIAMENT can also be considered institutional variables or conditions. The importance of them all was confirmed, although differently. While the importance of INSTITUTIONS was confirmed without any restrictions, political stability (POLSTAB) and democracy (DEMOCRACY/PARLIAMENT) showed clear path dependence, even if POLSTAB was also significant in the regression. Democracy was only part of a positive solution in Western developed states with rather little resources, when it was combined with political stability. In other contexts, it was rather the absence of democracy that seemed conducive to growth, on condition that political stability or good institutions were guaranteed. This does not necessarily refute the opinion of institutionalists who argue that democracy is good for growth due to a better institutional environment provided by these regimes. It only points to a number of exceptions: There are autocratic coutries with a good institutional environment, and it seems that the rents from oil and gas have helped to build these institutions.

Furthermore, political stability can also be detrimental if it is bought at a high cost, and factors of political instability vary in their impact on the economy. A possible study of these different kinds of political instability would require a more thorough qualification of the concept. Potential terrorist attacks targeting foreigners may be more of a deterrent than a government's repression of opposition movements. On the whole, however, the regression showed that political stability is good for the economy, and it seems that the GCC states are rather exceptional in this regard.

The Resource Curse Revisited

In my analyses, I also found evidence for resource curse effects, although these effects do not seem to be universal. In the regression, no significance of the different resource variables could be found. The assumption was that the impact of rents depended on the stage of development as well as on further circumstances such as the political regime. This was confirmed for the smaller samples.

The small sample (the GCC states) allowed for a distinction between extremely resource rich countries and modestly resource rich countries. While the first can expect to live from their resource rents, the latter cannot, and have to plan for a near future without oil and gas. Hence, the negative effects of natural resource abundance may not grow proportionally with the degree of natural resources wealth. They may only occur once enormous resource wealth makes any kind of productive economic activity unnecessary.

Interestingly, widely used resource dependence measures could be identified as reflecting heterogeneous situations which makes them less appropriate to test the resource curse hypothesis. While high reserves proved to be a good indicator for few incentives to diversify, RENT was certainly the most important condition, reflecting both Dutch disease effects and also incentives due to a habituation effect and grown rentier mentality.

The importance of the resource abundance variables/conditions for the GCC states confirms a suspicion from the theoretical discussion: The enormous wealth in natural resources distorts not only indicators, but also several widely assumed mechanisms of growth and development theory.

When it comes to Islam and Arab society, the regression could not give further insight. The percentage of Muslims (PERCENTMUSLIM) was, as expected, not significant. The attempt to introduce a dummy variable for Arab countries (ARAB), was bound to fail because features of collectivist societies are very common in many traditional societies, be they Arab or not. Partly, the importance of *wasta* may be reflected in the institutional variables/conditions.

Cultural issues were mainly discussed in the case studies. In the smaller sample, there was a lack of variance because all countries were Arab. In the enlarged fsQCA analysis, neither the percentage of Muslims in the population (PERCENTMUSLIM) nor the Arab "identity" of a country (ARAB) was an important condition.

The most important conclusion from the insignificance of these variables/conditions is the inappropriateness of such general operationalizations like the percentage of Muslims in the total population (PERCENTMUSLIM), or "Arabness". These operationalizations just test if there is a correlation between "Islam" or "Arabness" and growth while they have no informative value about the mechanism. From such a viewpoint, variable and outcome are linked by a black box that often contains a third, intervening variable. Arab societies may display certain obstacles to economic growth because they are collectivist. Collectivism is a very common feature in most less-developed economies. Therefore, the variable should be called "collectivist society" and not "Arab" or "African".

In about the same vein, Islamic economic institutions have been abandoned in many countries that are nonetheless predominantly Muslim. In most of them, Islam has never been of great importance for the economy. In the small sample of the GCC states and in the enlarged sample, it was possible to qualify the condition ISLAM more thoroughly based on an institutionalist approach, and here a high importance of Islamic economic institutions (ISLAM) correlated with a low rating for governance and regulatory quality – as well as other institutional indicators. Hence, the smaller samples offered the opportunity to base the variable on the assumed mechanisms between a cultural or religious feature and the economy. This analysis suggests that Islam and religiously motivated behavior is economically relevant (though overrated), even if the regression failed to prove this. On a larger scale, however, such a condition on religious institutions would require expert knowledge on many more countries and religions.

Outlook

Qualitative methods could indeed play a more important role in empirical studies on the factors determining economic growth. Where regressions have reached dead ends, they can take into account a more complex interplay of factors. This complexity, in turn, fits better into a changing field of economic studies, where regions and their specific, often path-dependent development trajectories attract attention. (Martin/Sunley 2006) What is more, not only regional economies but also regional centers of economic studies have mushroomed all over the world. Economic theory is no more the monopoly of Western universities but is now also shaped by contributions from India, China, Russia or South America. Also Islamic Economics is a field of study to be taken seriously.

Moreover, economics has broken away from what used to be called "political economy" because researchers wanted to assimilate it to the natural sciences. Today, however, the study of "political economy" has been revived. Researchers acknowledge processes too complicated or coincidential to be put into equations, and that universal theories based on unlikely assumptions may be of limited use. This study can be a starting point and it is hoped that slowly, data quality, operationalizations, and thereby my model can be improved.

Qualitative comparative methods, however, cannot replace detailed case studies: They can complement them by a comparative perspective, but the necessity of case studies will remain – mainly because of the complexity of issues but also because of the limits of measurement and operationalization. Comparable data will always entail a loss of information on the concrete situation, even more so when data quality is weak. Data problems were mostly due to the defaults of public accounting (Ramady 2013: 54): The exploitation of natural resources – an issue of enormous importance and public discussion today – is still not reflected properly in economic data and possibilities to amend this problem are limited due to weak data. Similarly, negative externalities are not accounted for: A huge problem in the GCC states is how construction, rapid population growth, consumption, and traffic put a heavy burden on the environment. What I have called non-oil growth in this study was bought at a high price, but mainly this price is excluded from calculations. It could only be mentioned in the case studies, and environmental issues in the GCC states will probably be objects for further study in the near future. However, studies, conferences, promises, and also the fact that the UAE could successfully attract the *International Renewable Energy Agency* will not suffice.

This price and also an increased dependence on the international economy actually put into question the entire process of economic diversification. The GCC states diversify their economies for a time when they run out of oil and gas, but they also assume that the international economy will follow the same principles. Many experts, however, agree that once fossil fuels will become scarce, the entire world economy will change. Many of them also assume that peak oil, which means the maximum of oil production, has already been transcended and that from now on production rates will decrease. (Heinberg 2003; Deffeyes 2010; Simmons 2005; Legett 2006; Legett 2013) Oil plays such an important part not only for energy and transport but also for the manufacturing of products, that a scarcity will be felt in all sectors.

The limits of the current diversification are more than obvious in the GCC states. Most segments of their economies remain either directly or indirectly dependent on fossil fuels. The extent of this ongoing dependency has recently also been demonstrated for the financial sector and stock market. (Al Kuwairi 2013) While it is obvious that an extreme reliance on natural resources poses a problem, it is easily overlooked that a normal, modest reliance on fossil fuels may also pose problems in the future. And this dependence will not be compensated for by diversification imitating industrialization processes from the West. Tourism, for example, relies on energy-intensive water desalination and cheap flights to the region. A scarcity of fossil fuels could make flights very expensive. Manufacturing and exports depend on international and regional demand, which, in turn, depend upon a sound growth performance of the international and national economies. If, however, a shortage of fossil fuels really leads to "degrowth" societies, as several researchers assume, this demand will decrease. There is still a fierce debate on the moment in future when oil will finally be depleted – if it will be depleted at all. The debate on how to respond to the threat of depletion is even more vigorous. The recent oil price highs, shale gas, as well as the environmental disasters that have resulted from drilling for new wells, have probably increased the number of adherents of degrowth or post-growth economists such as Serge Latouche (2006) or Herman Daly. Nonetheless, the real limit of this study seems to be a lack of knowledge on what will really happen if the world economy runs out of oil.

Appendices

Appendix A: Lists of Abbreviations

ADFEC	Abu Dhabi Future Energy Company
ADIA	Abu Dhabi Investment Authority
ADNOC	Abu Dhabi National Oil Company
ALBA	Aluminium Bahrain
AMD	Advanced Micro Devices
ARAMCO	Arabian American Oil Company (today: Saudi ARAMCO)
ASRY	Arab Shipbuilding and Repair Yard Company
BALEXCO	Bahrain Aluminium Extrusion Company
BANAGAS	Bahrain National Gas Company
BAPCO	Bahrain Petroleum Company
GCC	Gulf Cooperation Council
BP	British Petroleum
CDLR	Committee for the Defense of Legitimate Rights (Saudi Arabia)
CsQCA	Crisp Set Qualitative Comparative Analysis
DUCAB	Dubai Cable Company
DUBAL	Dubai Aluminium Company
DUGAS	Dubai Natural Gas Company
EADS	European Aeronautic Defense and Space Company
EMAL	Emirates Aluminium
FDI	Foreign Direct Investment

FERTIL	Ruwais Fertilizer Industries
FsQCA	Fuzzy set Qualitative Comparative Analysis
GARMCO	Gulf Aluminium Rolling Mill
GATT	General Agreement on Tariffs and Trade
GDP	Gross Domestic Product
GNI	Gross National Income
HDI	Human Development Index
IMF	International Monetary Fund
KAUST	King Abdallah University of Science and Technology
KIA	Kuwait Investment Authority
KOC	Kuwait Oil Company
LDC	Less-developed country
MIDAL	Middle East Aluminium Cable
MIRA	Movement for Islamic Reform in Arabia (Saudi Arabia)
NGO	Non-governmental Organization
NIE	New Institutional Economics
NUC	National Unity Committee (Bahrain)
OECD	Organisation for Economic Co-operation and Development
OPEC	Organization of the Petroleum Exporting Countries
OGC	Oman Gas Company
OPEC	Organization of Petroleum Exporting Countries
PDO	Petroleum Development Oman
PDQL	Petroleum Development Qatar, today Qatar Petroleum
QCA	Qualitative Comparative Analysis
QFC	Qatar Financial Centre
QFCA	Qatar Financial Centre Authority
QP	Qatar Petroleum
R&D	Research & Development

SABIC	Saudi Basic Industries Corporation
SAGIA	Saudi Arabian General Investment Authority
SAMA	Saudi Arabian Monetary Agency
SAMIRAD	Saudi Arabia Market Information
SWF	Sovereign Wealth Fund
TAKREER	Abu Dhabi Oil Refining Company
UAE	United Arab Emirates
UNESCO	United Nations Educational, Scientific and Cultural Organization
WTO	World Trade Organization
SOCAL/SoCal	Standard Oil of California

Abbreviation of Country Names

Country	Abbreviation
Abu Dhabi	ADH
Algeria	ALG
Bahrain	BAH
Botswana	BOT
Brazil	BRA
Brunei	BRU
Canada	CDN
Dubai	DUB
Ecuador	ECU
Egypt	EGT
Gabon	GAB
Indonesia	IDS
Iran	IRN
Iraq	IRQ
Kuwait	KWT
Malaysia	MAL

Mexico	MEX
Netherlands	NL
Nigeria	NGA
Norway	NOR
Oman	OM
Qatar	QAT
Saudi Arabia	SA
United Arab Emirates	UAE
United Kingdom	UK
United States	UAE
Venezuela	VEN

Appendix B: Description of Variables and Conditions

NDGROWTH/NDGROWTH

Growth rates of the non-oil part of the *per capita* GDP, calculated by subtracting the annual average resource rent (current prices, World Bank Green Accounting, corresponds to the variable RENT) *per capita* from the *per capita* GDP (current prices), taken from PENN World Table 6.3 chained by the US Dollar GDP deflator (as of the year 2000).

Downloads Rents: http://web.worldbank.org/WBSITE/EXTERNAL/TOPICS/ENVIRONMENT/EXTEEI/0,,contentMDK:20487830~menuPK:1187769~pagePK:148956~piPK:216618~theSitePK:408050,00.html

GDP and population data: http://pwt.econ.upenn.edu/php_site/pwt_index.php [15 January 2014].

LOGINVEST/INVESTMENT

Gross fixed capital accumulation/percentage of GDP. World Bank Development Indicators, data CD 2007. Enters the regression as natural logarithm. Calibration of the condition is not based on the natural log, but on raw data.

INVESTCAPITA

Condition based on the ratio of the annual average resource rent *per capita* and the annual per capita investments. Aims at an evaluation of the investment behavior.

LOGNDINITIAL
Initial non-oil GDP (PENN World Table 6.3; World Bank Green Accounting program for the resource rents, see above), enters the regression as natural logarithm.

FERTILITY/FERTILITY
Fertility rates taken from the World Bank Development Indicators http://data.worldbank.org/indicator [15 January 2014].

WOMENLAB/WOMENLAB
Percentage of women in the labor force taken from the World Bank Development Indicators http://data.worldbank.org/indicator [15 January 2014].

AVSCHOOLING
Average years of schooling of the population aged 15 and over, taken from the dataset used by Robert J. Barro and Jong-Wha Lee for their paper "International Data on Education Attainment: Updates and Implication", CIB Working Paper No. 42, 2000. Average of the data from 1970 to 2000 (data only available in five year intervals). Data for Oman, Qatar, the UAE and Saudi Arabia based on UNESCO and World Bank data.

Download of the dataset: http://www.cid.harvard.edu/ciddata/ciddata.html [16 January 2014].

UNESCO data: http://stats.uis.unesco.org/unesco/ReportFolders/ReportFolders.aspx?IF_ActivePath=P,50&IF_Language=eng [15 January 2015].

World Bank data: http://data.worldbank.org/indicator [15 January 2014].

HUMANCAP
Calibrated condition, based on AVSCHOOLING.

POLSTAB/POLSTAB
Perception of political stability taken from the World Bank governance indicators, averaged from 1996-2006. For the 25 cases of the large fsQCA sample, I estimated values from 1973 to 1995, based on the governance indicators methodology, qualitative information on the political history of these countries and the Polity IV index.

GOVREGULAR/GOVREGULAR
Average of two of the World Bank's governance indicators: governance and regulatory quality averaged from 1996 to 2006.

RULEOFLAW/RULEOFLAW

Average of two of the World Bank's governance indicators: rule of law and control of corruption averaged from 1996 to 2006.

INSTITUTIONS

Average of GOVREGULAR, POLSTAB and RULEOFLAW, used for the regression only.

DEMOCRACY/DEMOCRACY

The average of the civil liberties and political right (Freedom House), averaged from 1973 to 2005. Reversed in order to assign the high values to democratic countries.

ARAB/ARAB

Dummy variable/crisp set condition. Member of the Arab League (1), not member of the Arab League (0).

PERCENTMUSLIM/PERCENTMUSLIM

The percentage of Muslims in total population. Source: Robert Barro's "religion adherence data",

http://rbarro.com/data-sets/ [16 January 2014].

ISLAM

Condition based on six economic institutions which originate at least partly in Islamic law. A higher fuzzy value indicates a stronger role of Islamic institutions.

PARLIAMENT

Condition based on a different calibration of the Freedom House data. The turning point was set at 4.5, separating Kuwait, with the only strong parliament in the region, from the other GCC states.

RENT/RENT

Annual average resource rent per capita (1973-2005) in US Dollars (2000). Source: World Bank Environmental Accounting [16 January 2014].

RESERVES/RESERVES

Estimates of natural resource reserves, including oil, gas, coal/lignite and – if important resources –gold and silver was included. 2009 data was used (Source: BP).

RENTIER/RENTIERI

Natural resource rents as a percentage of government income, taken from the data set Michael Herb compiled for his paper "No Representation without Taxation? Rents, Development and Democracy," *Comparative Politics* 37, no. 3, April 2005, pp. 297-317, download from: http://www2.gsu.edu/~polmfh/pubs.htm [15 January 2014]; averaged from 1973 to last date available.

RENTIERII

Calibrated condition, based on the percentage of rents in government income, but calibrated differently: The turning point (0.5) is set between the extremely resource dependent and the resource dependent rentiers rather than between rentiers and semi-rentiers as in RENTIERI.

SXP/SXP

"Share of exports of primary products in GNP in 1970." Taken from the data set Sachs and Warner compiled for their paper "Natural Resource Abundance and Economic Growth", download from

http://www.cid.harvard.edu/ciddata/ciddata.html [16 January 2014].

Interaction Terms

INTERACTRENT

Interaction term of DEMOCRACY and RENT.

INTERACTRESERVES

Interaction term of DEMOCRACY and RESERVES.

INTERACTSXP

Interaction term of DEMOCRACY and SXP.

INTERACTRENTIER

Interaction term of DEMOCRACY and RENTIER.

List of Tables

List of Figures

Glossary of Arabic Terms

abaya (ʻabāya)
Long, usually black over-garment worn by women in the Gulf States.

ʻarabīya
Here: the Arabic language.

ʻaṣabīya
“Spirit of kinship”, kinship ties.

bid'a
"Innovation" which should be avoided.

Burj
"Tower" as in burj Dubai, burj al-Arab, burj Khalifa

dār al-ḥarb/dār al-islām
"House of war" standing for the land of the unbelievers as opposed to the "house" or the "land" of Islam.

dishdasha
Traditional long white garment worn by men in the Gulf States.

Emir, also: *amīr*
Ruler of an *Imarat (emirate)*.

falaj, pl. 'aflāj
Traditional irrigation system, mostly found in Oman.

falsafa
"Philosophy". Usually refers to the Islamic medieval philosophers.

fiqh
Islamic jurisprudence.

hajj
The pilgrimage to Mekka, one of the five pillars of Islam.

Hanbali (Ḥanbalīy)
Islamic school of law, mostly considered the strictest, going back to Ibn Ḥanbal.

Hegira (also Hejira or Hijrah)
The flight of Muhammad from Mecca to Medina in 622 A.D. The beginning of the Muslim era.

hijab (ḥijāb)
Veil worn by many Muslim women.

ḥiyal
Casuistry, gimmick to circumvent clauses of *sharia* law.

Ibāḍīyya
Islamic sect, most widespread in Oman.

ijmāʿ
Consensus of religious scholars as a source of sharia law.

ijtihād, ra'y al-ijtihād
Contentious method in Islamic jurisprudence: A Muslim scholar decides upon his/her own interpretation of the Koran and *sunna.*

Ikhwān
Brothers, brethren. Zealous bedouin warriors who aimed at spreading the Wahhabi faith.

Imamate
Entitiy ruled by an Imam (religious leader, mostly in a Shiite or Ibadi context).

intifāḍa
Literal meaning: "shaking off", mostly used for the Palestinian uprisings, but also for other uprisings in the Arab world.

kafīl
Sponsor, in the GCC states today: A local partner for a foreign business(man) who is required by law.

kalām
Theology (as opposed to *falsafa*, philosphy)

khalīj
The (Persian) Gulf.

Muḥammadīyya
Islamic movement founded by Ahmad Dahlan 1912 in Indonesia. Directed against idolatry and widespread synchretism in Indonesia.

majallah
Here: codex of civil law, mostly refers to the Ottoman Majallah.

majlis, pl. majālis
Assembly. Traditional meeting with the clan elders. Today also used for parliaments.

mujahideen (*mujāhidīn*)
"People doing *jihād*". Can be used very generally for anyone doing *jihād*, but is best known for the Muslim fighters against the Soviet invasion in Afghanistan.

murābaḥa
A double-sale instrument in Islamic banking. Usually, the bank purchases the product its clients would like to buy. It resells it to the client at a higher price.

mutawālī
Governor of a province or a *waqf*.

Muṭawwa'a
Zealous Wahhabists who worked for the spread of the Wahhabi doctrine and its strict observation. Closely linked to the Ikhwān. Today: Religious police in Saudi Arabia.

qādī
Islamic scholar, judge following Islamic law.

qiyās
"Analogy": A case is treated analogously to a similar, but not identical case in the Koran or *sunna*. Method of Islamic jurisprudence.

ribā
Depending on the interpretation interest or usury.

salaf, salafist
The forefathers: adherents of the faith of the forefathers (mostly in the sense of a "pure" Islam).

sharia (sharī'ah)
Islamic law contained in Koran and *sunna*.

sheikh
Notable, honorable person or tribal elder.

Shia
Islamic sect, adherents of 'Alī Abī Ṭālib, the fourth caliph, cousin and son-in-law of Muhammad. There are different branches, mostly depending on the number of Imams acknowledged.

shura (shūrah), maglis ash-shura
Advice, advisory council.

siyāsah
Politics, policy, realm of the ruler as opposed to the sharia, which was the realm of the religious scholars.

sunna
The living example and the daily life of Muhammad as a source of Islamic law additional to the Koran. Also: Sunni Muslims - largest group of Muslim believers, adherents of "*elected*" caliphs (Umayyads) as opposed to 'Alī who was related to Muhammad.

suq (Sūq)
Market, also stock market.

sura (sūrah)
A "chapter" of the Koran.

taqlīd
"Imitation" of previous verdicts of Islamic scholars as an instrument of Islamic jurisprudence.

ulema ('ulamā')
Muslim legal scholars, arbiters of *sharia* law.

umma ('ummah)
Community of (Muslim) believers.

uṣūl al-fiqh
Methodologies of Islamic jurisprudence (*fiqh*).

waqf, pl. awqāf
Religious endowment.

wasṭa
"Connection", medium; Mostly used in the sense of "using (family) connections to get a job or government service."

zakat (zakāt)
Islamic alms tax, one of the five pillars of Islamic faith.

Notes

Foreword

1 A frequently repeated and effective adagio, variously attributed to Sheikh Ahmed Zaki Yamani or to Robert Mabro

Chapter 1

1 In this book I will often refer to the states of the Gulf Cooperation Council as "Arab Gulf States", indicating that I mean only those states located at the Persian Gulf which consider themselves as Arab (and not as Persian), although Iraq is not included. There is an ongoing debate about the correct attribution of the Persian Gulf. Some claim that it has historically been called "Persian" Gulf, while others rightly adduce that the majority of the people living in the Gulf region are Arabs. By using the expression "Arab Gulf States", I do *not* intend to take the side of the latter group in this debate.

2 I will use non-oil growth and non-depletable growth interchangeably. Non-depletable growth was used by Stauffer and Lennox (1984) and is based on a sophisticated calculation of a non-depletable GDP. This calculation proved unfeasible and the notion, therefore, should not be mistaken for this method.

3 The failure of development economics and growth theory to explain these cases was observed earlier by Mostafa H. Nagi (1982), who highlighted the unusual circumstance of unlimited supply of capital in oil exporting economies.

4 Of course, Herman N. Daly's notion of a qualitative development goes much further than diversification and would probably not even include non-depletable growth as used in this study.

5 This must have caused Robert E. Lucas to boil down the difference between development theory and growth theory to this simple definition: "[...] we think of 'growth' and 'development' as distinct fields, with growth theory defined as those aspects of economic growth we have some understanding of, and development defined as those we don't." (Lucas 1988: 13)

6 The question of democracy and its correlation with growth has especially had many contradictory answers. (Quinn/Woolley 1995) The same holds true for other widely used variables such as trade policy/openness or foreign direct investment. Sometimes, the significant correlation stems from one very important case. (Stiglitz 2005)

7 Defined as "socially devised constraints on individual action", (Douglass North, quoted in Clague 1997: 17).

8 Exogenous growth theory was particularly noted in the 1950s by Robert M. Solow (1956). Later on, the exogeneity of technical improvements was put into question and modified in endogenous growth models. (Romer 1990)

9 An important part of this literature is focused on Islamic finance (Noorzoy 1982), but in the second half of the 20th century, general literature on Islamic Economics also boomed. (Ragab 1980; al Haq 1996; Khan 1985; Chapra 1991; Choudhury/Malik 1992; Philipp 1990; Nienhaus 1982; id. 2006; Kuran 1986; id. 1995; id. 1996; id. 1997; id. 2004; id. 2005; id. 2008; Maurer 2001)

10 In this study, the notions non-oil growth and non-depletable growth will be used interchangeably. Diversification, however, could also take place without growth and should be viewed as distinct.

11 Sectorial data is more widely available. How the single sectors grew in the GCC states can be taken from Koren/Tenreyro 2010.
12 One of his first works *The Comparative Method. Moving beyond Qualitative and Quantitative Strategies* (Ragin 1987) was followed by many others, elaborating the method.
13 It can be easily demonstrated why fuzzy sets are usually preferable to crisp sets in social science. Countries can only rarely be defined as "democratic" or "non-democratic". There are many countries in between, being neither perfectly democratic nor totally undemocratic. The same holds true for most variables. (Ragin 2000)
14 An interesting compilation of links to growth data sets can be found on http://www.bris.ac.uk/Depts/Economics/Growth/datasets.htm [last access 28 December 2013].
15 Unfortunately, the world value survey has not collected data from the GCC states save for very recent data on Saudi Arabia. (World Value Survey 2011)

Chapter 2

1 The Sovereign Wealth Fund Institute lists 74 SWFs, most of which have their origin in commodities. Among the GCC States, there is more than one SWF per country.
2 Monika Mühlböck reported cases of women in high positions who have personal assistants to do most of the work. Interview with author, January 2010.
3 The strict surveillance of public morals in several Gulf countries may also have deterring effects on potential investors and on tourism. Tourists prefer to walk freely, to dress as they like and to drink alcohol without fearing punishment. In most GCC states, this is possible. Saudi Arabia, on the contrary, dispenses with "normal" tourism and sticks to its strict view of public morals – much to the benefit of neighboring Bahrain, where Saudis go if they want more freedom.
The same is true of Islamic terrorism: Even if most Muslims despise terrorism and deem it unIslamic, terrorist activities by Muslims cannot be viewed as disconnected from a fundamentalist interpretation of Islam. Both investors and tourists will rather avoid countries with a high threat of terrorism. However, this threat can hardly be assessed in advance, and today terrorist activity can occur anywhere.
4 Davidson (2008: 195) reports allegations of government officials buying-off terrorist groups.
5 Today, the assets of the KIA are estimated at around US $386 billion. (Soverign Wealth Found Institute 2014) Kuwait has important investments in the Daimler AG and Citigroup (bailout of 6 million), but there are also investments in Kuwait and in the Arab world. The National Technology Enterprises Company, created in 2002, can be duly called a Sovereign Wealth Enterprise. The SWF Institute assigned KIA a rating of 6 on a scale from 0 (not at all transparent) to 10 (perfectly transparent).
6 Data on Defense Spending can be obtained from the Stockholm International Peace Research Institute (SIPRI 2019). In constant US $ (2005), Kuwait's military expenditure rose from 3,154 million to 12,168 million in1990 and 15,857 million in 1991.
7 The Project Kuwait was launched in 1997. Information at that time could be obtained from www.kuwaitproject.com [October 10, 2010]. Meanwhile, this website has ceased to exist and information on the "Project Kuwait" can hardly be found anymore. In 2005, the Middle East Economic Digest (16 September 2005, Vol. 49, Issue 37) reported about the opposition which "described the project as the theft of the century" (MEED 2005: 9) since concessions should go to foreign companies.
8 In 2003, there was even a negative FDI inflow to Kuwait compared to positive inflows into all other GCC countries. (ESCWA 2009: 3; ESCWA 2008: 6)
9 Interview with Monika Fatima Mühlböck, Vienna, January 2010.
10 Most of them are US-American schools (Texas A&M, Carnegie Mellon, Georgetown, Weill Cornell Medical College...), but also the Dutch Stenden University.
11 To get an idea of expatriates' experiences in Qatar, the website www.qatarsucks.com is of informative.
12 Traditionally, land was used by Bedouins, nomads and a few settled farmers according to their own rules. The Public Lands Distribution Ordinance (1968) broke up the traditional agricultural structures

by allocating land. The new landholders needed to fulfill basic requirements (25 percent of the land had to be developed within a certain time) in order to obtain ownership.

13 According to the ESCWA, FDI in Saudi Arabia in 2003 was about US $778 million. This is about as much as for Oman ($640 million), but much less than for the United Arab Emirates ($4,292 million). Given the size of Saudi Arabia and its population, this is comparatively little. From 2003 to 2004, however, FDI in Saudi Arabia rose dramatically. ESCWA sees the reasons for this in the opening of important sectors to FDI which had hitherto been reserved for nationals. (ESCWA 2009: 3; ESCWA 2008: 6f)

14 Between 1982 and 2000, only one new domestic bank was granted a license, while there was no new foreign bank at all on the Saudi market. After 2000, other GCC banks opened branches in Saudi Arabia. (Hamidy 2006: 9).

15 This may also have an irritating effect. The documentation of the admission process to the WTO reveals that *zakat* was the object of several questions by the working party. (WTO 2005: 9f)

16 The text has been published in the *Arab Law Quarterly*, 16, 1, 2001, pp. 101-104. For a discussion see Niblock/Malik 2007: 188-199.

17 Given that Saudi Arabia was among the reformers of the year in 2008, there is also a valuable special report on the reforms. (Belayachi/Haidar 2008)

18 With "merely" 70 billion US dollars, Dubai's Investment Corporation is far behind Abu Dhabi's Soverign Wealth Funds. (SWFI 2014)

19 The Central Bank of the UAE publishes a statistical bulletin (Monthly Banking & Monetary Statistical Bulletin), including historic trade statistics.

20 The Cooperation Agreement was signed in 1988 and can be read on the website of the European Union External Action (2014).

Chapter 3

1 An impressive list of 62 variables which have so far been used in recent cross-country regressions on growth is given in Obersteiner/Wilk 1999: 23.

2 For example, Keynes pointed to the fact that goods are not sold at once and that cash is not spent at once. Both goods and cash, therefore, are nonproductive for a certain time. In neoclassical models, it is usually assumed that there are no delays between production, sale, and the withdrawal of money and its spending. Almost every reaction takes place immediately.

3 On this point, Keynes contradicts the neoclassical view that future events and developments could be predicted with probability calculations.

4 It is widely accepted to speak of a single model, even if Harrod and Domar's starting points were different. While Domar explains the impact of capital and production on demand, Harrod tries to explain how demand influences savings and investments. Harrod's and Domar's models are separately explained in Bretschger 1999: 14-16 (Domar) and 6-20 (Harrod).

5 Diminishing returns to capital mean that adding any unit of capital augments the output less than the unit added before.

6 It is clear that the assumptions of the Harrod-Domar had to be eased: Solow allowed for a variable capital coefficient: labor could be replaced by capital. In addition, Solow introduced the assumption of diminishing marginal returns of labor and capital separately and constant returns to scale for capital and labor together.

7 In "The Origins of Endogenous Growth" (1994), Romer compares assumptions and implications of different endogenous models and tries to summarize them in one.

8 Other assumptions are: full employment, identical production functions for both countries, no transportation costs, and identical demand.

9 Rosenstein-Rodan, born in Krakow, had personal experience in less-developed countries. For his first articles on the Big Push see Rosenstein-Rodan 1943 and 1944.

10 Rostow tried to give not only a theory of development, but a theory of history in general. Writing like Marx on history and development, he proposed a totally different view on history and certainly aimed to replace him. Hence the "Non-Communist Manifesto" in his title. (Preston 1996: 175).

11 There are several ways to categorize institutions. Some, like Douglass North (1990: 3f; 1991: 97), distinguish between formal and informal institutions. Also, the suggestion to differentiate between external and internal institutions on the one hand and between the way they are sanctioned, if at all, is made. (Voigt 2003: 413).

12 As far as I know, only Preston uses this term (Preston 1996: 196).

13 "Without a theory they had nothing to pass on except a mass of descriptive material waiting for a theory, or a fire." (Klein 2000: 457)

14 *'Aṣabīya* can be translated as "spirit of kinship" as it can mostly be found in tribal societies.

15 This is just one example for the sample-oriented approach of regressions. As I concentrate on a small sample of cases, the result is not useful for my study. It even bears the risk of wrong inferences. Additionally, Vicente's regression suffers from a lack of data, which he admits himself. Only in rare cases does the lack of data make an author really cautious about his results. Authors often mention data problems at the beginning of their paper, but fail to take it into account in their conclusion. (Dezhbakhsh 2005).

16 Another point seems to be quite logical: Given that oil was discovered when only few countries were industrialized, there were not many countries where oil could have been exported to.

17 A good definition of "rent" is provided by Richards and Waterbury (2008: 16f).

18 Marshall contrasts the concept of the rent with that of the *quasi* rent, which may also be obtained from the use of a machine. (Marshall 1997: 74)

19 Gilbar (1997: xv) speaks of producer governments and consumer governments.

20 Despite his focus on other issues, Foley uses the history of oil, oil companies, and oil prices as a structure of his study. (Foley 2010)

21 Kuran discusses why true profit and risk sharing instruments have failed to become more widespread despite being favored by religious scholars. On the contrary, *murābaḥa* is much more popular and still "serves as a cumbersome form of interest" (Kuran 2004: 45).

22 I will use the two notions interchangeably.

23 In the interviews I conducted with expatriate workers in the GCC states, this point was confirmed by all respondents.

24 Interview with Monika Fatima Mühlböck, Vienna, January 20, 2010. For the potential of women in economic life, especially skilled private sector jobs, see the chapter "If only women will work" in Foley (2010) and also Ramady (2013: 259f).

Chapter 4

1 Also Schneider and Wagemann count among their "good practices" the use of other methods for the analysis of the research question. (Schneider/Wagemann 2007: 267).

2 In my description of fsQCA, I follow closely the recent book by Schneider/Wagemann 2007. A recent and comprehensive English introduction to the different kinds of QCA (csQCA, fsQCA and mvQCA) is given in Rihoux/Ragin 2008.

3 Readers might be puzzled by the qualification of Ragin's method as "qualitative" in spite of the use of data. Sometimes, "qualitative" is replaced by "case-oriented", which might be a more appropriate description. Ragin, however, considers his method as belonging to the qualitative branch of empirical research. (Schneider/Wagemann 2007: 27)

4 Ragin prefers "condition" to "variable" because the latter is usually associated with regressions. In this study, there will be both regressions and QCA-analyses, using partly the same "variables". Depending on the context, I will differentiate between the notions, but in some cases, I refer to the variable itself,

not depending on the kind of analysis it is used in. Within the analysis, conditions will be in small caps, while variables will appear in capital letters.

5 There are different possibilities to name conditions and to indicate that they are present or absent. One possibility is to use capital letters for present conditions and lower-case letters for absent conditions. Referring to Boolean algebra, also the sign "¬" could precede a condition to indicate its absence. In this study, I have opted to use th swung dash o indicate that a condition is absent (~A means that condition A is absent) The "and" of formal logic is indicated by a star "*", while the logical "or" is given by "+". Thus, when I write A*B + ~A*B → Y, this means that there are two sufficient combinations of conditions leading to Y: A and B combined or the absence of A combines with the presence of B. Of course, this can be written more easily: In fact, in this case, B is the sufficient conditions for Y: B →Y.

6 It is also common practice to use capital letters for present, and lower-case letters for absent, conditions. Here, a difference is made between between variables (in capital letters) and conditions (small caps), hence the swung dash was the chosen.

7 Neal Caren and Aren Panofsky (2005) developed a new method dealing with different notions of time in QCA. So far, this method is only a step into the direction of adding temporality to QCA and serious problems occur once it is applied to fuzzy sets.

8 Coverage can also be very high if the condition A is not consistent. It is therefore important that consistency is calculated first. If a condition is not consistent, coverage is misleading.

9 While Libya would have also been an interesting case for this sample, data was not available.

10 In this study, I make no difference between the notions of "non-oil" and "non-depletable" growth even if non-oil growth might hide the fact that rents from other natural resources are also included. Oil-independent or rent-independent, however, will not be used.

11 Also called Hotelling's rule or scarcity rent. The concept goes back to the US-American statistician Harold Hotelling (1895-1973).

12 Stauffer and Lennox (1984) take into account another phenomenon which has somehow disappeared in recent literature on resource rents. The rentier multiplier, like the Keynesian multiplier, has the effect of increasing a country's GDP by more than just by the oil rents because the oil rents are spent by the government, for example, as salaries or pensions and are spent again: "Even though the 'non-depletable' GDP is better than the usual GDP, the 'oil-independent GDP', is less, often very much less, than the 'non-depletable' GDP and provides a better measure of the economy's 'non-oil' performance and its degree of dependence upon its oil rents." (Stauffer/Lennox 1984: 29) Elsewhere, Stauffer estimates that the rentier multiplier can be assumed to be between 1.3 and 1.6, which means, that the rent increases further by 30 to 60 percent, which makes the non-oil GDP even smaller. (Stauffer 1987: 34)

13 Unfortunately, the GDP in current prices is called "real GDP" in the PENN World Table, because it has been adjusted for purchasing power parity. (Heston/Summers/Aten 2009)

14 A considerable part, between 18 to 30 percent according to Byman and Green (1999: 21), of the income from oil seems to disappear in the private accounts of the members of the ruling families.

15 This variable is called "growth rate of real GDP *per capita* (constant prices: chain series)", (Heston/Summers/Aten 2009).

16 Please note that the software calibration avoids the fuzzy values 0 and 1.

17 Depending on the country and the available health care, the replacement rate is actually between 2.1 and 2.33.

18 Sachs and Warner tested four variables. The first three all measure resource dependence: SXP (natural resource exports divided by GDP in 1970), SNR (mineral production divided by NP in 1970) and PXI70 (natural resource exports divided by total exports in 1970). The last one (LAND) is the log of arable land area divided by population, has the lowest coefficient (-0.39 compared to -8.28 for SXP), and is the only variable not based on dependence on natural resources. It seems that they have chosen the variable with the highest coefficients and the best t-statistics, which happened to be SXP (Sachs/Warner 1995a/1997: table III, p. 39).

19 Contrary to Sachs and Warner (and also to other authors such as Richard Auty), I will not include agricultural products because they are only partly a gift from nature: They are primary products but not natural resources, since they have to been grown and cared for, and they are not depletable (only the soil is to a certain extent). From the GCC perspective, more agricultural production could be regarded as diversification, but irrigation problems and climate make agriculture very costly. They are still net importers of agricultural products.
20 The data has been taken from Robert Barro's "Religion Adherence Data", downloaded free from http://rbarro.com/data-sets/ [15 January 2014].
21 Meanwhile, the Polity IV index has expanded and covers almost all sample countries apart from Brunei. (Polity IV Project/Center for Systemic Peace 2013)
22 Complete data, standard deviations included, can be downloaded from http://info.worldbank.org/govehttp://info.worldbank.org/governance/wgi/index.aspx#home [15 January 2014].

Chapter 5

1 For the qualitative basis of these estimates see the country cards of Abu Dhabi and Dubai in chapter two.
2 Interestingly, speeding fines as well as an indirect tax collected via a toll system are an important source of income. In 2009, this kind of income is estimated to have been 1.5 times higher than income from fossil fuel production. (McGinley 2010)
3 Kuwait started development early, too, but invested much more abroad than locally.
4 A comprehensive list of data quality indicators can be taken from "Data Appendix for a Space-Time System of National Accounts" on the Penn World Table website.
Penn World Table 6.1 (PWT 6.1), http://pwt.econ.upenn.edu/php_site/pwt_index.php [15 Janaury 2014]: 14-16.
5 The GCC states are all classified as traditional monarchies, while Kuwait was once classified as "occupied", and Bahrain, Kuwait and Oman were also classed as no-party monarchies. Meanwhile, there is a dataset ("The Authoritarian Regime Dataset") available on http://www.svet.lu.se/ARD/ [12 August 2012].

Chapter 6

1 From 1998 to 2002, Qatar had the highest non-oil growth rates according to my calculations and also to the calculations of Fasano and Iqbal (2003).

Bibliography

Aarts, Paul; Gerd Nonneman (2005), *Saudi Arabia in the Balance: Political Economy, Society, Foreign Affairs*, London; Hurst 2005.

Adams, Richard H. (2006), *Migration, Remittances and Development: The Critical Nexus in the Middle East and North Africa*, United Nations Expert Group Meeting on International Migration and Development in the Arab Region.

Adelman, Irma; Cynthia Taft Morris (1973), *Economic Growth and Social Equity in Developing Countries*, Stanford: Stanford University Press.

Aggarwal, Rajesh; Tarik Yousef (2000), "Islamic Banking and Investment Financing", *Journal of Money*, Credit and Banking, 32, 1, pp. 93-120.

Aghion, Philippe (1998), *Endogenous Economic Growth*, Cambridge, MA.: MIT Press.

Alkhozai, Mohammed (2008), "The Challenges of Vocational Training: BIBF, an Institute of Higher Education", in: Davidson; Mackenzie Smith (eds.), *Higher Education in the Gulf States, Shaping Economies, Politics and Culture*, (SOAS Middle East Issues), London, Berkeley, Beirut: Saqi, pp. 144-148.

Al-Kuwari, Duha (2013), "Mission impossible? Genuine economic development in the Gulf Cooperation Council Countries", *Kuwait Programme on Development, Governance and Globalisation in the Gulf States, Research Paper Series* No. 33.

Al-Zumai, Fahad (2013), "Kuwait's Political Impasse and Rent-Seeking Behaviour:

A Call for Institutional Reform", *Kuwait Programme on Development, Governance and Globalisation in the Gulf States, Research Paper Series* No. 29.

Amin, S. H. (1991), *Legal System of Kuwait* (Middle East Legal System Series), Glasgow: Royston Publishers.

Amīn, Samīr (1976), *Unequal Development: An Essay on the Social Formations of Peripheral Capitalism,* transl. by Brian Pearce, Hassocks: Harvester Press.

Amuzegar, Jahangir (2001), *Managing the Oil Wealth. OPEC's Windfalls and Pitfalls*, London; New York: I.B. Tauris.

Annen, Kurt (2003), "Kulturelle Ökonomik am Beispiel von Sozialkapital", in: Blümle/Goldschmidt/Klump a. o. (eds.), *Perspektiven einer kulturellen Ökonomik* (Kulturelle Ökonomik Bd. 1), Münster: LIT Verlag, pp. 29-44.

Ansary, Abdullah F. (2008), "A Brief Overview of the Saudi Arabian Legal System", *Globalex*, New York University, http://www.nyulawglobal.org/globalex/saudi_arabia.htm [15 January 2014].

Arab News (2010), "Economic Experts Call for Overhaul of Zakat System", *Arab News*, July 9, 2.

http://www.arabnews.com/node/349878 [15 January 2014].

Arrow, Kenneth J. (1962), "The Economic Implications of Learning by Doing", *Review of Economic Studies*, 29, pp. 155-173.

Audretsch, David B.; Max C. Keilback; Erik E. Lehmann (2006), *Entrepreneurship and Economic Growth*, Oxford: Oxford University Press.

Auty, Richard M. ed. (2001), *Resource Abundance and Economic Development,* (WIDER studies in Development Economics), Oxford: Oxford University Press.

Auty, Richard M. (1990), *Resource-Based Industrialization: Sowing the Oil in Eight Developing Countries*, Oxford: Clarendon Press.

Auty, Richard M. (1993), *Sustaining Development in Mineral Economies: The Resource Curse Thesis*, London: Routledge 1993.

Auty, Richard M. (2001), "A Growth Collapse with High Rent Point Resources: Saudi Arabia", in: Id. (ed.), *Resource Abundance and Economic Development*, Oxford: Oxford University Press for UNU-Wider, pp. 193-207.

Auty, Richard M. (2006), "Patterns of Rent-Extraction and Deployment in Developing Countries. Implications for Governance, Economic Policy and Performance", *UNU-WIDER Research Paper*, 16.

Azzam, Henry T., *Souk Al-Manakh Stock Market Crises in Kuwait*, United Gulf Bank 1983.

Baba, Stephen A. (1997), "Democracy and Inefficiency", *Economics and Politics*, 9, pp. 99-114.

Badran, Hania (1995), *Sozio-kulturelle Aspekte der Wirtschaftsbeziehungen im Arabischen Raum*, Linz: Rudolf Trauner.

Baghat, Gawdat (1999), "Education in the Gulf Monarchies: Retrospect and Prospect", *International Review of Education/Internationale Zeitschrift für Erziehungswissenschaft/Revue Internationale de l'Education*, 45, 2, pp. 127-136.

Bahadir, Sefik A. (1990), "Wirtschaftliche Integration der Arabischen Golfstaaten: Probleme und Perspektiven des Golf-Kooperationsrates", in: Pawelka, Peter; Maho A. Aves (eds.), *Arabische Golfstaaten in der Krise*, Frankfurt a. Main, pp. 118-147.

Bahrain Economic Development Board (2008), *Our Vision. From Regional Pioneer to Global Contender. The Economic Vision 2030 for Bahrain*, download from http://www.bahrainedb.com/en/about/Pages/economic%20vision%202030.aspx#.Uo4S1eK2qbU [16 January 2014].

Baldwin, Robert E. (1966), *Economic Development and Growth*, New York; London; Sydney: John Wiley.

Ballantyne, William M. (1980), *Legal Development in Arabia. A Selection of Articles and Addresses on the Arabian Gulf*, London: Graham & Trotman Publishers.

Ballantyne, William M. (1986), *Commercial Law in the Arab Middle East: The Gulf States*, London: Lloyd's of London.

Ballantyne, William M. (2006), "Commercial Law: The Conflict in Shari'a and Secular Law Public Policy", in Behdad/Nomani (eds.), *Islam and the Everyday World. Public Policy Dilemmas*, London; New York: Routledge, pp. 141-164.

Barro, Robert J. (1997), *Determinants of Economic Growth. A Cross-Country Empirical Study*, Cambridge MA; London: The MIT Press.

Barro, Robert J.; Xavier Sala-i-Martin (2004), *Economic Growth*, Cambridge, MA; London: The MIT Press.

Basedau, Matthias; Wolfram Lacher (2006), "A Paradox of Plenty? Rent Distribution and Political Stability in Oil States", *GIGA Working Paper*, 21.

BBC (1999), "World: Middle East Election Defeat for Qatar Women", *BBC News*, 9 March, on http://news.bbc.co.uk/2/hi/middle_east/293171.stm [15 January 2014].

BBC (2009), "Dubai $10bn help from Abu Dhabi 'was Loan, not Handout'", *BBC News*, December 17 http://news.bbc.co.uk/2/hi/8418013.stm [15 January 2014].

Beblawi, Hazem; Giacomo Luciani eds. (1987), *The Rentier State (Nation, State and Integration in the Arab World Vol. II)*, London; New York; Sydney: Croom Helm.

Beblawi, Hazem (1987), "The Rentier State in the Arab World", in: Id., Giacomo Luciani (eds.), *The Rentier State (Nation, State and Integration in the Arab World Vol. II)*, London; New York; Sydney: Croom Helm, pp. 49-62.

Beck, Martin; Oliver Schlumberger (1999), "Der Vordere Orient – ein entwicklungspolitischer Sonderfall? ", in: Pawelka, Peter; Hans-Georg Wehling (eds.), *Der Vordere Orient an der Schwelle zum 21. Jahrhundert*, Wiesbaden: Verlag für Sozialwissenschaften, pp. 57-80.

Beck, Martin (2007), "Der Rentierstaats-Ansatz und das Problem abweichender Fälle", *Zeitschrift für Internationale Beziehungen*, 14, 1, Juni 2007, pp. 43-70.

Benthall, Jonathan (2002), "Organized Charity in the Arab-Islamic World", in: Hastings Donnan (ed.), *Interpreting Islam*, London; Thousand Oakes: Sage, pp. 150-166.

Berg, Werner (2003), "Die unerbittlichen Folgen des Vergleichs: Die Entdeckung der Kultur als wichtige Determinante des ökonomischen Prozesses", in: Blümle/Goldschmidt/Klump a. o. (eds.), *Perspektiven einer kulturellen Ökonomik* (Kulturelle Ökonomik Bd. 1), Münster: LIT Verlag, pp. 45-64.

Berger, Ron; Avi Silbiger; Ram Herstein; Bradley R. Barnes (2014), "Analyzing business-to-business relationships in an Arab context", *Journal of World Business*, 8.

Beutel, Jörg (2012), "Conceptual Problems of Measuring Economic Diversification, as Apllied to the GCC Economics", in: Luciani, Giacomo (ed.), *Resources Blessed. Diversification and the Gulf Development Model,* Berlin: Gerlach Press, pp. 29-70.

Birks, J.S.; C.A. Sinclair (1979), "Preparations for Income after Oil: Bahrain's Example", *British Journal of Middle Eastern Studies,* 6, 1, pp. 39-57.

Birks, J. S.; C.A. Sinclair (1987), "Successful Education and Human Resource Development - the Key to Sustained Economic Growth", in: B.R. Pridham (ed.), *Oman: Economic, Social and Strategic Developments,* pp. 145-167.

Borchert, Manfred (1975), "Das Heckscher-Ohlin-Theorem", *Wirtschaftswissenschaftliches Studium*, 4, 3, p.141.

Bretschger, Lucas (1999), *Growth Theory and Sustainable Development*, Cheltenham; Northampton, MA: Edward Elgar 1999.

Brinitzer, Ron (2003), *Religion – eine institutionenökonomische Analyse* (Religion in der Gesellschaft, Bd. 14), Würzburg: Ergon Verlag.

Bristol-Rhys, Jane (2008), "The Dilemma of Gender-Separated Higher Education in the United Arab Emirates", in: Davidson; Mackenzie Smith (eds.), *Higher Education in the Gulf States, Shaping Economies, Politics and Culture,* (SOAS Middle East Issues), London, Berkeley, Beirut: Saqi, pp. 98-109.

British Petroleum (2013), *Statistical Review of World Energy 2013,* http://www.bp.com/en/global/corporate/about-bp/energy-economics/statistical-review-of-world-energy-2013.html [7 January 2014].

Brown, Nathan J. (1997), *The Rule of Law in the Arab World: Courts in Egypt and the Gulf,* Cambridge; Cambridge University Press.

Brunnschweiler, Christa N. (2008), "Cursing the Blessing? Natural Resource Abundance, Institutions and Economic Growth", *World Development*, 36, 3, pp. 399-419.

Bulliet, Richard W. (1983), "Review of Bernard Lewis' *The Muslim Discovery of* Europe", *The American Historical Review*, 88, 2, pp. 439-440.

Butt, Gerald (2001), "Oil and Gas in the UAE", in: Ibrahim al Abed; Peter Hellyer (eds.), *United Arab Emirates. A new perspective*, London: Trident Press, pp. 231-248.

Byman, Daniel L.; Jerrold D. Green (1999), "The Enigma of Political Stability in the Persian Gulf Monarchies", *Middle East Review of International Affaris*, 3, 20.

Caren, Neal; Aren Panofsky (2005), "TQCA: A Technique for Adding Temporality to Qualitative Comparative Analysis", *Sociological Methods and Research,* 34, pp. 147-172.

Center for Systemic Peace, *Political Regime Characteristics and Transitions, 1800-2012 (Polity IV Project),* http://www.systemicpeace.org/inscr/inscr.htm [15 January 2015].

Central Bank of Kuwait (2007), *Kuwait 2020,* Kuwait City: Cental Bank of Kuwait.

Chapra, M. Umer (1991), "The Need for a New Economic System", *Review of Islamic Economics*, 1, 1, 1991, pp. 9-47.

Charles, Lorraine (2015), "The Perception of Young Emiratis on Emiratisation", in: Annika Kropf and Mohamed A. Ramady (eds.), *Employment and Career Motivation in the Arab Gulf States: The Rentier Mentality Revisited*, Berlin: Gerlach Press (forthcoming).

Chaudhry, Kiren Aziz (1994), "Economic Liberalization and the Lineages of the Rentier State", *Comparative Politics* 27, 1, pp. 1-26.

Chaudhry, Kiren Aziz (1997), *The Price of Wealth. Economics and Institutions in the Middle East*, Ithaca; London: Cornell University Press (Cornell Studies in Political Economy).

Choudhury, Masudul Alam; Malik, Uzir Abdul (1992), *The Foundations of Islamic Political Economy*, Basingstoke: Macmillan.

Clague, Christopher (1997), "The New Institutional Economics and Economic Development", in: id. (ed), *Institutions and Economic Development. Growth and Governance in Less-Developed and Post-Socialist Countries*, Baltimore; London: The John Hopkins University Press, pp. 13-36.

Clague, Christopher; Philip Keefer a. o. (1997), "Institutions and Economic Performance: Property Rights and Contract Enforcement", in: Clague (ed.), *Institutions and Economic Development. Growth and Governance in Less-Developed and Post-Socialist Countries*, Baltimore; London: The John Hopkins University Press, pp. 67-90.

Clague, Christopher; Philip Keefer a. o. (1997a), "Democracy, Autocracy, and the Institutions Supportive of Economic Growth", in: Clague (ed.), *Institutions and Economic Development. Growth and Governance in Less-Developed and Post-Socialist Countries*, Baltimore; London: The John Hopkins University Press, pp. 91-120.

Crystal, Jill (1995), *Oil and Politics in the Gulf. Rulers and Merchants in Kuwait and Qatar,* Cambridge: Cambridge University Press updated edition (Cambridge Middle East Library 24).

Coase, Ronald H. (1984), "The New Institutional Economics", *Journal of Institutional ad Theoretical Economics*, 140, pp. 229-232.

Cole, Simon (1992), *The Gulf States: A Business Handbook*, Oxford: Blackwell.

Colton, Nora (2011), "Social Stratification in the Gulf Cooperation Council States", *Kuwait Programme on Development, Governance and Globalisation in the Gulf States, Research Paper Series* No. 14.

Cordon, W. Max; J. Peter Neary (1982), "Booming Sector and De-Industrialisation in a Small Open Economy", *The Economic Journal*, 92, 368, pp. 825-848.

Coulson, Noel J. (1984), *Commercial Law in the Gulf States: The Islamic Legal Tradition*, London: Graham & Trotman.

Daly, Herman E. (1990), "Boundless Bull", *Gannett Center Journal*, 4, 3, pp. 113.

Daly, Herman E. (1999), "Globalization versus internationalization – some implications", *Ecological Economics,* 31, pp. 31-37.

Davidson, Christopher M. (2006), "After Shaikh Zayed: The Politics of Succession in Abu Dhabi and the UAE", Middle East Policy, 13, 1, pp. 42-59.

Davidson, Christopher M. (2008), *Dubai. The Vulnerability of Success*, London: Hurst & Company.

Davidson, Christopher; Peter Mackenzie Smith eds. (2008), *Higher Education in the Gulf States, Shaping Economies, Politics and Culture* (SOAS Middle East Issues), London, Berkeley, Beirut: Saqi.

Davidson, Christopher M. (2009), *Abu Dhabi. Oil and Beyond*, New York: Columbia University Press 2009.

Deffeyes, Kenneth S., *When Oil Peaked,* New York, NY: Hill and Wang 2010.

DeLong-Bas, Natana J. (2008), *Wahhabi Islam. From Revival and Reform to Global Jihad*, Oxford: Offord University Press.

Dezhbakhsh, Hashem (2005), "The Political Economy of Statistical Evidence: Economic Data Problems in Developing Countries and their Impact on Empirical Research", in: Ayogu, Melvin; Don Ross (eds.), *Development Dilemmas. The Methods and Political Ethics of Growth Policy*, London; New York: Routledge (Routledge Studies in Development Economics 41), pp. 80-102.

Dietz, Simon; Eric Neumayer; Indra de Soysa (2007), "Corruption, the Resource Curse and Genuine Saving", *Environmental and Development Economics* 12, pp. 33-53.

Domar, Evsey .D. (1946), "Capital Expansion, Rate of Growth and Employment", *Econometrica* 14, pp. 137-147.

DPA, "Bahrain's Income Tax, First in Gulf, Sparks Opposition", *Khaleej Times,* June 26, 2007.

Drummond, Helga; F.B. Al-Anazi (1997). "Leadership Styles in Saudi Arabia: Public and Private Sector Organisations Compared", *Cross Cultural Management: An International Journal 4* (1997), pp. 3-8.

Dutton, R.W. (1987), "Updating Agriculture and Associated Rural Entreprises", in: B.R. Pridham (ed.), *Oman: Economic, Social and Strategic Developments*, pp. 94-117.

Easterly, William (2001), *The Elusive Quest for Growth. Economists' Adventures and Misadventures in the Tropics*, Cambridge, MA; London: MIT.

Eisenstadt, Shmuel N. (1973), *Traditional Patrimonialism and Modern Neopatrimonialism.* Beverly Hills: Sage Publications.

El-Kharouf, Farouk; Sulayman Al-Qudsi; Shifa Obeid (2010), "The Gulf Corporation Council Sovereign Wealth Funds: Are They Instruments for Economic Diversification or Political Tools?", *Asian Economic Papers*, Winter/Spring, Vol. 9, No. 1, pp. 124-151.

Erdmann, Gero; Ulf Engel (2006), "Neopatrimonialism Revisited – Beyond a Catch-All Concept", *GIGA Research Program: Legitimacy and Efficieny of Political Systems*, 16, www.giga-hamburg.de/content/publikationen/pdf/wp16_erdmann-engel.pdf (A) [15 January 2014].

Esanov, Akram; Martin Raiser; Willem Buiter (2001), "Nature's Blessing or Nature's Curse: The Political Wconomy in Transition of Resource-Based Economies", *Working Paper of EBRD*, 66, www.nbp.pl/konferencje/radisson/Mowcy/buiter/Buiter_paper2.pdf [15 January 2014].

ESCWA (2001), *Economic Diversification in the Oil-Producing Countries: The Case of the Gulf Cooperation Council Economies*. New York: United Nations, ESCWA.

ESCWA (2008), *Foreign Direct Investment Report*, ESCWA, 3 September, Technial Paper 1.

ESCWA (2009), *Foreign Direct Investment Report*, ESCWA, 2 September, Technial Paper 2.

Etherington, Richard (2010), "Review: Norway – Oil Production Past its Peak but not its Best", *OilEdge*, 6 Nov. http://www.oiledge.com/n/Review_Norway_oil_production_past_its_peak_but_not_its_best/f1d0f716.aspx [18 January 2014].

European Union External Action (2014), *EU relations with the Gulf Cooperation Council*, EU External Action website http://eeas.europa.eu/gulf_cooperation/index_en.htm [15 January 2014].

Evans, Peter (1995), *Embedded Autonomy: States and Industrial Transformation,* Princeton: Princeton University Press.

Evans-Pritchard, Ambrose (2008), "Dubai Debt Risk as Crisis Rocks Gulf Banks", *The Telegraph*, October 13, http://www.telegraph.co.uk/finance/globalbusiness/3191538/Dubai-debt-risk-as-crisis-rocks-Gulf-banks.html [15 January 2014].

Executive Council (2007), *Highlights: Dubai Strategic Plan (2015)*. Dubai: Executive Office.

Fargues, Philippe (2013), "International Migration and the Nation State in Arab Countries", *Middle East Law and Governance*, 5, pp. 5-35.

Fasano, Ugo; Zubair Iqbal (2003), *GCC Countries: From Oil Dependence to Diversification*, IMF Publication, http://www.imf.org/external/pubs/ft/med/2003/eng/fasano/index.htm#1 [17 January 2014].

Fitch, Asa (2010), "The Souk al-Manakh: Kuwait's first big bubble", *The National*, March 13.

Flick, Uwe (2004), *Triangulation. Eine Einführung*, Wiesbaden: Verlag für Sozialwissenschaften 2004.

Flörkemeier, Holger (2003), "Kulturelle Vielfalt, Transaktionskosten und Aussenhandel", in: Blümle/Goldschmidt/Klump a. o. (eds.), *Perspektiven einer kulturellen Ökonomik* (Kulturelle Ökonomik Bd. 1), Münster: LIT Verlag, pp. 95-108.

Foley, Sean (2010), *The Arab Gulf States. Beyond Oil and Islam*. Boulder; London: Lynne Rienner.

Frank, André Gunder (1969), *Latin America: Underdevelopment or Revolution - Essays on the Development of Underdevelopment and Immediate Enemy,* New York: Monthly Review Press.

Frank, André Gunder (1971), *Sociology of Development and Underdevelopment of Sociology,* London: Pluto Press.

Frenkel, Michael; Hans-Rimbert Hemmer (1999), *Grundlagen der Wachstumstheorie*, München: Wahlen.

Fromherz, Allen J. (2011), *Qatar. A Modern History*, London: I.B. Tauris.

Fürtig, Henner (2006), "Saudi-Arabiens WTO-Beitritt: Bestätigung des wirtschaftlichen Reformkurses", *GIGA-Focus*, 5.

Gabriel, Tom (1990), "Social Analysis and Rural Development Strategies for Oman", *Arid Soil Research and Rehabiliation*, 4, 1, pp. 75-83.

Garrett, Larry Neal; Ali Farghali (1987), "On the Need for an Integrated Policy Infrastructure for the Delivery of Technical and Vocational Education in the Arab Gulf Region", *Comparative Education*, 23, 3, pp. 317-328.

Gause, F. Gregory III. (1994), *Oil Monarchies. Domestic and Security Challenges in the Arab Gulf States*, New York: Council on Foreign Relations Press.

Geddes, Barbara (1999), "What do we Know about Democratization after Twenty Years?", *Annual Review of Political Science,* 2, pp. 115-144.

Gelb, Alan et al (1988), *Oil Windfalls: Blessing or Curse?*, Oxford; New York: Oxford University Press for the World Bank 1988.

Gellner, Ernest (1992), *Der Islam als Gesellschaftsordnung*, Stuttgart 1992.

General Secretariat for Development Planning (2008), *Qatar National Vision 2030*, http://www.gsdp.gov.qa/portal/page/portal/gsdp_en/qatar_national_vision/qnv_2030_document [15 January 2014].

Gerakis, A.S.; O. Roncesvalles (1983), "Bahrain's Offshore Banking Center", *Economic Development and Cultural Change*, 31, 2, pp. 271-293.

German Trade and Invest (2009), "Wirtschaftstrends Jahreswechsel 2013/14 - Kuwait", http://www.gtai.de/GTAI/Navigation/DE/Trade/maerkte,did=937490.html [15. January 2014].

Ghanem, Shihab M. (2001), "Industrialization in the UAE", in: Ibrahim al Abed; Peter Hellyer (eds.), *United Arab Emirates. A new perspective*, London: Trident Press, pp. 260-276.

Gibert, Bastien; Axel Maraut; Benjamin Telle (2005), *Enjeux et Perspectives pour les Emirats Arabes Unis. Et après le Pétrole?*, (Entreprise et Management), Paris: Harmattan.

Gilbar, Gad G. (1997), *The Middle East Oil Decade and Beyond: Essays in Political Economy*, London a. o.: Cass.

Global Footprint Network (2010), *National Ecological Footprint and Biocapacity for 2007. Result from National Footprint Acconts 2010 edition*, http://www.footprintnetwork.org/en/index.php/GFN/page/footprint_data_and_results [15 January 2014].

Grandmaison, Colette le Cour (1984), "L'Eau du Vendredi. Droits d'Eau et Hierarchie Social en Sharqiya (Sultanate of Oman)", *Études Rurales*, 94, pp. 7-42.

Greif, Avner (1994), "Cultural Beliefs and the Organization of Society: A Historical and Theoretical Reflection on Collectivist and Individualist Societies", *Journal of Political Economy*, 102, 5, pp. 912-950.

Griffin, K. (1999), "Culture and Economic Growth: A General Argument with Illustrations from the Islamic World", *Journal of Islamic Studies, 10,* 2, pp. 126-141.

Grigorian, David; Vlad Manole (2005), "A Cross-Country Nonparametric Analysis of Bahrain's Banking System", *IMF Working Paper,* 05, 117, pp.1-16.

GSDP (General Secretariat for Development Planning), (2008), *Qatar National Vision 2030*, Doha: GSDP.

Guiso, Luigi; Paola Sapienza; Luigi Zingales (2003), "People's Opium? Religion and Economic Attitudes", *Journal of Monetary Economics*, vol. 50, 1, pp. 225-282.

Gylfason, Thorvaldur; T. Herbertsson; Gylfi Zoega (1999), "A Mixed Blessing: Natural Resources and Economic Growth", *Macroeconomic Dynamics*, 3, pp. 204-225.

Gylfason, Thorvaldur (2001), "Natural Resources, Education, and Economic Development", *European Economic Review*, 45, 4-6, pp. 847-859.

Gylfason, Thorvaldur (2001a), "Lessons from the Dutch Disease: Causes, Treatment, and Cures", *Institute of Economic Studies, Working Paper Series*. http://www.ioes.hi.is/publications/wp/w0106.pdf [14 January 2014].

Gylfason, Thorvaldur (2006), "The Dutch Disease: Lessons from Norway", report for the Trinidad Tobago Chamber of Commerce and Industry, 1, p. 4. http://notendur.hi.is/gylfason/Trinidad2006.pdf [15 January 2014].

Gylfason, Thorvaldur (2007), "The International Economics of Natural Resources and Growth", *Minerals and Energy*, https://notendur.hi.is//~gylfason/biblio.htm [8 October 2014].

Hadenius, Axel; Jan Teorell (2004), "Same, Same - But Different. Assessing Alternative Indices of Democracy", Prepared for delivery at the 2004 Annual Meeting of the American Political Science Association.

Hadenius, Axel; Jan Teorell (2007), "Pathways from Authoritarianism", *Journal of Democracy* 18, 1, pp. 143-156.

Hallaq, Wael B (1984), "Was the Gate of Ijtihad Closed?", *International Journal of Middle East Studies*, 16, 1, pp. 3-41.

Hallaq, Wael B (1997), *A History of Islamic Legal Theories: An Introduction to Sunnī Uṣūl Al-Fiqh*, New York: Cambridge University Press.

Hamdan, Sara; Travis Pantin (2009), "Divisions over Riyadh as GCC Central Bank Site", *The National,* May 6.

Hamaiza, Adel (2015), "Cross-National Heterogeneity of Hofstede's 'Arab Countries'", in: Annika Kropf and Mohamed A. Ramady (eds.), *Employment and Career Motivation in the Arab Gulf States: The Rentier Mentality Revisited*, Berlin: Gerlach Press (forthcoming).

Al Hamidy, Abdulrahman (2006), "Banking Sector Issues in Saudi Arabia", *BIS Papers,* 28.

Hanushek, Eric A.; Dennis D. Kimko (2000), "Schooling, Labor-Force Quality and the Growth of Nations", *The American Economic Review*, 90, 5, pp. 1184-1208.

Al Haq, Irfan (1996), *Economic Doctrines of Islam: A Study in the Doctrines of Islam and their Implications for Poverty, Employment, and Economic Growth*. Herndon: International Institute of Islamic Thought.

Harrod, Roy F. (1939), "An Essay in Dynamic Theory", *Economic Journal,* 49, pp. 14-33.

Harriss, John ed. (1997), *The New Institutional Economics and Third World Development*, London: Routledge.

Hassan, Hussein (2011), *Contracts in Islamic Law* (Library of Islamic Law Vol. 1), Basingstoke: I.B. Tauris.

Hecht-El Minshawi, Béatrice (2007), *Wirtschaftswunder in der Wüste: Strategien für langfristigen Erfolg in den Golfstaaten*, Heidelberg: Redline Wirtschaft.

Hein, Wolfgang (2000), "Andre Gunder Frank (1929 -), Metropolen, Satelliten und das Weltsystem", *Entwicklung + Zusammenarbeit*, 3, pp. 80-83.

Heinberg, Richard (2003), *The Party's Over: Oil, War and the Fate of Industrial Societies*, Gabriola: New Society Publishers.

Henni, Abdelghani (2012), "Kuwait 2012 downstream industry profile", *Arabian Oil and Gas*, February 26. http://www.arabianoilandgas.com/article-9991-kuwait-2012-downstream-industry-profile/ [02 October 2014].

Herb, Michael (1999), *All in the Family: Absolutism, Revolution, and Democracy in the Middle Eastern Monarchies*, Albany, NY: State University of New York Press, (SUNY Series in Middle Eastern Studies).

Herb, Michael (2002), "Does Rentierism Prevent Democracy?", *Annual Meeting of the American Political Science Associatio*n, August 29 – September 1.

Herb, Michael (2005), "No Representation without Taxation? Rents, Development, and Democracy", *Comparative Politics*, 37, 3, pp. 297-316.

Herb, Michael (2009), "A Nation of Cureaucrats: Political Participation and Economic Diversification in Kuwait and the United Arab Emirates", *International Journal of Middle East Studies,* 41 (3), pp. 375-395.

Hertog, Steffen (2005), "Segmented Clientelism: The Political Economy of Saudi Economic Reform Efforts", in: Aarts/Nonneman (eds.), *Saudi Arabia in the Balance: Political Economy, Society, Foreign Affairs*, London; Hurst, pp. 111-143.

Hertog, Steffen (2008), "Two-level negotiations in a fragmented system: Saudi Arabia's WTO accession", *Review of international political economy*, 15 (4). pp. 650-679.

Hertog, Steffen (2010), *Princes, Brokers, and Bureaucrats. Oil and the State in Saudi Arabia*. Ithaca: Cornell University Press.

Hertog, Steffen (2012), "A comparative assessment of labor market nationalization policies in the GCC", in: id. (ed.), *National employment, migration and education in the GCC, The Gulf Region: Economic development and diversification*, 4, Berlin: Gerlach Press.

Hertog, Steffen (2012a), "Public Industry as Tool of Rentier Economy Diversification: The GCC Case", in: Luciani, Giacomo (ed.), *Resources Blessed. Diversification and the Gulf Development Model,* Berlin: Gerlach Press, pp. 115-138.

Hertog, Steffen (2013), "The private sector and reform in the Gulf Cooperation Council", *Kuwait Programme on Development, Governance and Globalisation in the Gulf States, Research Paper Series* No. 30.

Herzog, Karen ed. (1998), *Investitionsführer Naher Osten. Ägypten, Bahrain, Irak, Iran, Israel, Jemen, Jordanien, Kuwait, Libanon, Oman, Qatar, Saudi-Arabien, Syrien, Vereinigte Arabische Emirate,* Frankfurt am Main: FAZ.

Heston, Alan; Robert Summers; Bettina Aten (2009), *Penn World Table Version 6.3,* Center for International Comparisons of Production, Income and Prices at the University of Pennsylvania.

Hirschman, Albert O. (1958), *The Strategy of Economic Development*, New Haven.

Hirschman, Albert O. (1995), *A Propensitiy to Self-Subversion*, Cambridge, MA: Harvard University Press.

Hodgins, Peter; Mark Beswetherick (2009), *Insurance in the Kingdom of Saudi Arabia,* Clyde & Co, July 2009.

Hofstede, Geert (1980), *Culture's Consequences: International Differences in Work-related Values* Sage, Beverly Hills, CA: Sage.

Howard, Philip N. (2010), *The Digital Origins of Dictatorship and Democracy. Information Technology and Political Islam* (Oxford Studies in Digital Politics), Oxford: Oxford University Press.

Howard, Philip N.; Muzammil M. Hussain (2013*), Democracy's Fourth Wave? Digital Media and the Arab Spring*, Oxford/New York: Oxford University Press.

Howitt, Peter and David N. Weil. (2008), "Economic growth." The New Palgrave Dictionary of Economics. Second Edition. Eds. Steven N. Durlauf and Lawrence E. Blume. Palgrave Macmillan.

Hsu, Tanya Cariina (2015), "The Anthropological Elements of Failed Saudisation: Historicism, Image, Islam and Tribe", in: Annika Kropf and Mohamed A. Ramady (eds.), *Employment and Career Motivation in the Arab Gulf States: The Rentier Mentality Revisited*, Berlin: Gerlach Press (forthcoming).

Huff, Toby E.; Wolfgang Schluchter eds. (1999), *Max Weber & Islam*, New Brunswick; New Jersey: Transaction Publishers.

Human Rights Watch Middle East (1997), *Routine Abuse, Routine Denial. Civil Rights and the Political Crisis in Bahrain*, New York; Washington; London; Brussels: Human Rights Watch. http://www.hrw.org/legacy/reports/1997/bahrain/ [15 January 2014].

Hvidt, Martin (2013), "Economic diversification in GCC countries: Past record and future trends", *Kuwait Programme on Development, Governance and Globalisation in the Gulf States/LSE*, Working Paper No. 27.

IFAD (1999), *Assessment of Rural Poverty in West and Central Africa*, Rome, August.

IMF (2006), "Saudi Arabia: Financial System Stability Assessment, including Reports on the Observance of Standards and Codes on the following topics, Monetary and Financial Policy Transparency, Banking Supervision, and Payment Systems," *IMF Country Report*, 6/199, June.

IRIN/UN (2009), "Botswana: Less development revenue as diamonds lose their shine", *IRIN Humanitarian News and Analysis*, 3 April, http://www.irinnews.org/report/83784/botswana-less-development-revenue-as-diamonds-lose-their-shine [16 January 2015]

Jones, Alan (2005), "Ibāḍiyya", in: Lindsay Jones (ed.), *Encyclopaedia of Religion*, 6, Detroit; New York: Macmillan/Thomson/Gale, pp. 4247-4249.

Jones, David (2015), "Key Drivers for Employee Engagement and Productivity of GCC Nationals and Expatriates: A Longitudinal Study 2010 – 2012", unpublished paper delivered at the 5th Gulf Research Meeting, Cambridge, August 24th -28th.

Karl, Terry Lynn (1997), *The Paradox of Plenty: Oil Booms and Petro-States*, Berkely, CA.

Kassem, Sami; Ghazi M. Habib (1989), *Strategic management of Services in the Arab Gulf States: Company and Industry Cases*, Berlin a. o.: De Gruyter.

Katz, Mark N. (2004), "Assessing the Political Stability of Oman", *Middle East Review of International Affairs*, 8, 3, pp. 1-10.

Kaufmann, Daniel; Aart Kraay; Massimo Mastruzzi (2010), "The Worldwide Governance Indicators: Methodology and Analytical Issues", *World Bank Policy Research Working Paper* 5430.

Kechichian, Joseph A. (1985), "The Gulf Cooperation Council: Search for Security", *Third World Quaterly*, 7, 4, pp. 853-881.

Keech, William R. (1995), *Economic Politics: The Costs of Democracy,* New York: Cambridge University Press.

Keynes, John Maynard (1930), *Treatise on Money*, New York: Harcourt Brace.

Keynes, John Maynard (1936), *The General Theory of Employment, Interest and Money*, Cambridge: Macmillan Cambridge University Press.

Ibn Khaldûn (1967), *The Muqaddima*, translated by Franz Rosenthal; N.J. Dawood (ed.), Princeton: Princeton University Press.

Khaleej Times/DPA (2007), "Bahrain's income tax, first in Guld, sparks opposition", *Khaleej Times*, 26 June. http://www.khaleejtimes.com/DisplayArticleNew.asp?xfile=data/business/2007/June/business_June839.xml&section=business&col= [7 January 2014].

Khan, M. A. (1985), "Economic Growth and Development in Islam", *The Search*, 6, pp. 142-166.

Kincaid, Harold (2005), "Development Theory and the Philosophy of Science", in: Ayogu, Melvin; Don Ross (eds.), *Development Dilemmas. The Methods and Political Ethics of Growth Policy* (Routledge Studies in Development Economics 41), London; New York: Routledge, pp. 62-79.

Koch, Christian (2014), "Constructing a viable EU-GCC partnership", *Kuwait Programme on Development, Governance and Globalistion in the Gulf States, Research Paper Series* No. 34.

Klein, Peter G. (2000), "New Institutionalist Economics", in: Boudewijn, Bouckeart; Gerrit De Geest (eds.), *Encyclopedia of Law and Economics, Vol. I: The History and Methodology of Law and Economics*, Cheltenham: Edward Elgar, pp. 456-489.

Koppers, Simon (1995), *Economic Analysis of the Gulf Cooperation Council (GCC)* (European University Studies Bd. 1783), Frankfurt am Main; Berlin: Peter Lang.

Koren, Miklós; Silvana Tenreyro (2010), "Volatility, Diversification and Development in the Gulf Cooperation Council Countries", *Kuwait Programme on Development, Governance and Globalisation in the Gulf States, Research Paper Series No. 9*, London: London School of Economics.

Knack, Stephen; Philipp Keefer (1995), "Institutions and Economic Performance: Cross-Country Tests Using Alternative Institutional Measures", *Economics and Politics*, 7, pp. 207-227.

Kolb, Joachim (2015), "The Two Universe of Entrepreneurship in the UAE: An Answer to the Gulf Employment conundrum", in: Annika Kropf and Mohamed A. Ramady (eds.), *Employment and Career Motivation in the Arab Gulf States: The Rentier Mentality Revisited*, Berlin: Gerlach Press (forthcoming).

Kormendi, Roger C.; Philip C. Mcguire (1985), "Macroeconomic Determinants of Growth: Cross Country Evidence", *Journal of Monetary Economics*, 16, pp. 141-163.

Krommer, Andreas (1986), *Entwicklungsstrategien der arabischen Golfstaaten (Bahrain, Kuwait, Oman, Qatar, Saudi-Arabien, Vereinigte Arabische Emirate): Eine vergleichenden Analyse im Lichte der relevanten Ansätze der Theorie der Entwicklungspolitik* (Europäische Hochschulschriften: Reihe 5, Volks- und Betriebswirtschaft, 728), Frankfurt a. Main a. o.: Lang.

Kropf, Annika (2010), "Resource Abundance vs. Resource Dependence in Cross-Country Growth Regressions", *OPEC Energy Review*, 34, 2, pp. 107-130.

Kubursi, Atif A. (1984), *Oil, Industrialization & Development in the Arab Gulf States*, London a. o.: Croom Helm 1984.

Kuran, Timur (1986), "The Economic System in Contemporary Islamic Thought: Interpretation and Assessment", *International Journal of Middle East Studies*, 18, pp. 136-164.

Kuran, Timur (1995), *Private Truth, Public Lies*, Cambridge, MA; London: Harvard University Press.

Kuran, Timur (1997), "Islam and Underdevelopment: An Old Puzzle Revisited", *Journal of Institutional and Theoretical Economics*, 153, pp. 41-71.

Kuran, Timur (2001), "The Provision of Public Goods under Islamic Law: Origins, Impact, and Limitations of the Waqf System", *Law & Society Review*, 35, 4, pp. 841-898.

Kuran, Timur (2004), *Islam & Mammon. The Economic Predicaments of Islamism*, Princeton; Oxford: Princeton University Press.

Kuran, Timur (2005), "The Absence of the Corporation in Islamic Law", *The American Journal of Comparative Law*, 53, 4, pp. 785-834.

Kuran, Timur (2008), "The Scale of Entrepreneurship in Middle Eastern History: Inhibitive Roles of Islamic Institutions", *ERID Working Paper Number 10*.

Kuwait News (2010), "Blair's 'Kuwait Vision 2035' report proposes solutions to key issues", 15 March. http://www.kuna.net.kw/ArticleDetails.aspx?id=2068419&language=en [17 January 2014].

Kvande, Halvor (2012), "The GCC Aluminum Industry – The Development of Aluminum Production Capacity in the Middle East," in: Luciani, Giacomo (ed), *Resources Blessed. Diversification and the Gulf Development Model*, Berlin: Gerlach Press, pp. 213-233.

Lacroix, Stephane (2014), "Secularism in an Islamic State: the case of Saudi Arabia", *Kuwait Programme on Development, Governance and Globalisation in the Gulf States, Research Paper Series*, London School of Economics.

Latouche, Serge (2006), *Le pari de la décroisance,* Paris: Fayard.

Legett, Jeremy (2006), *Half Gone. Oil, Gas, Hot Air and the Global Energy Crisis*, London: Portobello.

Legett, Jeremy (2013), *The Energy of Nations: Risk Blindness and the Road to Renaissance*, New York: Routledge.

Legrenzi, Matteo (2008), "Did the GCC Make a Difference? Institutional Realities and (Un)Intended Consequences", in: C. Harders and M. Legrenzi (eds.), *Beyond Regionalism? Regional Cooperation, Regionalism and Regionalization in the Middle East*. Aldershot: Ashgate, pp. 107–24.

Leite, Carlos.; Jens Weidmann (1999), "Does Mother Nature Corrupt? Natural Resources, Corruption, and Economic Growth", *IMF Working Paper* 99/85.

Lepenies, Philipp; Manfred Nitsch (2000), "John Maynard Keynes (1883 – 1946). Unsichere Zukunftserwartungen als Motor und Bremse ökonomischer Entwicklung", *E+Z - Entwicklung und Zusammenarbeit*, 12, pp. 354- 356.

Lewicki, T. (2009), "al- Ibāḍiyya" *Encyclopaedia of Islam, Second Edition*, P. Bearman, Th. Bianquis, C.E. Bosworth, E. van Donzel, W.P. Heinrichs (eds.), Brill, 2009. Brill Online. Freie Universitaet Berlin. <http://www.brillonline.nl/subscriber/entry?entry=islam_COM-0307> [October 05, 2009]

Lewin, Tamar (2008), "American Universities Find a Global Market", *The New York Times*, February 18.

Lewis, Bernard (1982), *The Muslim Discovery of Europe*, New York: W. W. Norton.

Lewis, Bernard (1993), *Islam and the West*, New York: Oxford University Press.

Lewis, Bernard (2002), *What went Wrong? The Clash of Islam and Modernity in the Middle East*, Oxford: Oxford University Press.

Lipset, Seymour Martin (1959), "Some Social Requisites of Democracy: Economic Development and Political Legitimacy", *The American Political Science Review*, 53, 1, pp. 69-105.

Looney, Robert E. (1990), *Economic Development in Saudi Arabia: Consequences of the Oil Price Decline*, Greenwich, CT; London: Jai Press.

Looney, Robert E. (1991), "Diversification in a Small Oil Exporting Economy. The Impact of the Dutch Disease on Kuwait's Industrialization", *Resource Policy*, pp. 31-41.

Looney, Robert E. (1992), "Real or Illusory Growth in an Oil-Based Economy: Government Expenditures and Private Investment in Saudi-Arabia", *World Development,* 20, 9, pp.1367-1375.

Looney, Robert E. (1994), *Industrial Development and Diversification of the Arabian Gulf*
Economies. Greenwich, CT: JAI Press.

Looney, Robert E. (2004), "The Saudi Arabian Quandary: The Economy's Inability to Sustain Growth", *ODTÜ Gelişme Dergisi*, 31 (Haziran), 2004, pp. 1-22.

Lucas, Robert E. Jr. (1988), "On the Mechanics of Economic Development", *Journal of Monetary Economics*, 22, pp. 3-42.

Luciani, Giacomo (1987), "Allocation vs. Production States: A Theoretical Framework", in: Beblawi/Luciani (eds.), *The Rentier State (Nation, State and Integration in the Arab World Vol. II)*, London; New York; Sydney: Croom Helm, pp. 63-82.

Luciani, Giacomo (2012), "GCC Refining and PEtrocehmical Sectors in Global Perspective", in: id. (ed.), *Resources Blessed. Diversification and the Gulf Development Model,* Berlin: Gerlach Press, pp. 183-212.

Luciani, Giacomo (2005), "From Private Sector to National Bourgeoisie: Saudi Arabian Business", in: Aarts/ Nonneman (eds.), *Saudi Arabia in the Balance: Political Economy, Society, Foreign Affairs*, London; Hurst, pp. 144-181.

Lulu, Tahiyya, "The Real Story of Bahrain's Divided Society", *The Guardian*, March 3, 2011.

Mahoney, James; Gary Goertz (2006), "A Tale of Two Cultures: Contrasting Quantitative and Qualitative Research", *Political Analysis*, 14, pp. 227-249.

Maneschi, Andrea (1998), *Comparative Advantage in International Trade. A Historical Perspective*, Cheltenham; Northampton: Edward Elgar.

Marshall, Alfred (1997), *Principles of Economics*, New York: Prometheus Books 1997.

Martin, Ron; Peter Sunley (2006), "Path Dependence and Regional Economic Evolution", *Journal of Economic Geography*, Vol. 6, 4, pp. 395-437.

Masum Billah, Mohammed (1993), "Life Insurance? An Islamic View", *Arab Law Quaterly*, Vol. 8, No. 4, pp. 315-324.

Maurer, Bill (2001), "Engineering an Islamic Future: Speculations on Islamic Financial Alternatives", *Anthropolgy Today*, 17, 1, pp. 8-11.

Mauro, Paolo (1995), "Corruption and Growth", *The Quarterly Journal of Economics,* Vol. 110, No. 3, pp. 681-712.

Mc Dowall, Angus (2013), "Saudi Arabia, angered over Mideast, declines Security Council seat", *Reuters*, 18 October.

McGinley, Shane (2010), "Salik and Speed Fines are Vital Government Revenue", *Arabian Business,* September 25, http://www.arabianbusiness.com/salik-speed-fines-are-vital-government-revenue-economist-351033.html [15 January, 2014].

McKinsey & Company (2007), "GCC Education Breakout: Preparing GCC Youth with the Skills to Meet Job Market Needs", *GCC Education Leaders Conference*, November 28-29.

MEED (2005), "MPs step up anti-Project Kuwait rhetoric", *Middle East Economic Digest*, September 16.

MEED (2007), "Muscat Says No to Islamic Banking: Oman Ignores Growth in Sharia Sector", *MEED Middle East Economic Digest*, 16 February.

MEED, (2011), "Oman spends ot quelle unrest", *MEED Middle East Economic Digest,* 29 July.

Meur, Gisèle de; Benoît Rihoux; Sakura Yamasaki (2008), "Addressing the Critiques of QCA", in: Rihoux, Benoît; Charles C. Ragin, *Configurational Comparative Methods. Qualitative Comparative Analysis (QCA) and Related Techniques* (Applied Social Research Methods Series, Vol. 51), Los Angeles; London a. o.: Sage, pp. 147-166.

Mills, Robin, (2012), "Pioneering Bahrain finds itself reliant on Saudi largesse", *The National* (UAE), May 2, http://www.thenational.ae/business/industry-insights/energy/pioneering-bahrain-finds-itself-reliant-on-saudi-largesse [01 October 2014].

MoEP (Ministry of Economy and Planning) (2004), *Long-Term Strategy for the Saudi Economy*: Riyadh: Ministry of Economy and Planning, Kingdom of Saudi Arabia, www.mep.gov.sa/index.jsp;jsessionid=E8C97863827DC1926339E242A3B10C04.alfa?event=ArticleView&Article.ObjectID=53 [15 December 2013].

MONE (Ministry of National Economy) (1995), *Long-Term Development Strategy (1996–2020): Vision for Oman's Economy – 2020.* Sultanate of Oman, Ministry of National Economy, www.mone.gov.om/viewPublication.aspx?id=366, accessed 24 January 2012.

Mueller, Hans-Georg, "GCC Industrial Development", in: Luciani, Giacomo (ed.), *Resources Blessed. Diversification and the Gulf Development Model,* Berlin: Gerlach Press, pp. 139-182.

Al-Muhairi, Butti Sultan Butti Ali (1996), "The Development of the UAE Legal System and Unification with the Judicial System", *Arab Law Quarterly*, 11, 2, pp. 116-160.

Mühlböck, Monika (2007), "Zur gesellschaftlichen Stellung der Frau in den arabischen Golfstaaten", *ORIENT,* IV, pp. 44-51.

Murphy, Kevin M.; Andrei Shleifer, Robert W. Vishny (1993), "Why is Rent-Seeking so Costly to Growth?", *The American Economic Review,* 83, 2, Papers and Processing of the Hundred and Fifth Annual Meeting of the American Economic Association, pp. 409-414.

Myrdal, Gunnar (1957), *Economic Theory and Underdeveloped Regions*, London: Duckworth.

Myrdal, Gunnar (1970), *The Challenge of World Poverty*, London: Allen Lane.

Nagi, Mostafa H. (1982), "Development with Unlimited Supplies of Capital: The Case of OPEC", *The Developing Economies*, pp. 3-20.

Al-Nasr, Tofol (2009), "Critique of Michael L. Ross' Article, 'Oil, Islam, and Women'", *New Voices in Public Policy*, Georges Mason University, 4.

Naughton, Shahnaz; Tony Naughton (2000), "Religion, Ethics and Stock Trading: The Case of an Islamic Equities Market", *Journal of Business Ethics*, 23, 2, pp. 145-159.

Neumayer, Eric (2004), "Does the 'Resource Curse' hold for Growth in Genuine Income as well?", *World Development*, 32, 10, pp. 1627-1640.

Niblock, Tim; Monica Malik (2007), *The Political Economy of Saudi Arabia,* London; New York: Routledge.

Nienhaus, Volker (1982), *Islam und moderne Wirtschaft. Positionen, Probleme und Perspektiven*, (Islam und Westliche Welt, 6), Graz; Wien; Köln: Styria Verlag 1982.

Nienhaus, Volker (2006), "Zakat, Taxes, and Public Finance in Islam", in: Behdad/Nomani (eds.), *Islam and the Everyday World. Public Policy Dilemmas*, London; New York: Routledge 2006, pp. 165-192.

Niethammer, Katja (2006), "Voices in Parliament, Debates in Majalis, and Banners on Streets: Avenues of Political Participation in Bahrain", *EUI Working Papers*, RSCAS 27.

Nizar Hamzeh, A. (1994), "Qatar: The Duality of the Legal System", *Middle Eastern Studies*, 30, 1, 1994, pp. 79-90.

Noorzoy, M. Siddieq (1982), "Islamic Laws on Riba (Interest) and their Economic Implications", *International Journal of Middle East Studies* 14, pp. 3-17.

Nordhaus, William D. (1975), "The Political Business Cycle", *Review of Economic Studies,* 42, pp. 169-190.

North, Douglass C. (1989), "Institutions and Economic Growth: A Historical Introduction", World Development, 17, pp. 1319-1332.

North, Douglass C. (1990), *Institutions, Institutional Change and Economic Performance*, Cambridge: Cambridge University Press.

North, Douglass C. (1991), "Institutions", *Journal of Economic Perspectives*, 5, 1, pp. 97-112.

Noueihed, Lin; Frederik Richter (2011), "Saudi Sends Troops, Bahrain Shi'ites Call it 'War'", *Reuters*, March 14 http://www.reuters.com/article/2011/03/14/us-bahrain-protests-forces-idUSLDE72D0KH20110314 [22 December 2013].

Nutzinger, Hans G.; Stephan Panther (2003), "Homo oeconomicus vs. Homo culturalis: Kultur als Herausforderung der Ökonomik", in: Blümle/Goldschmidt/Klump a. o. (eds.), Perspktiven einer kulturellen Ökonomik (Kulturelle Ökonomik Bd. 1), Münster: LIT Verlag, pp. 287-310.

Oakeshott, Michael (1958/1993), *Morality and Politics in Modern Europe: The Harvard Lectures,* S. R. Letwin (ed.), New Haven, CT: Yale University Press.

Obersteiner, Michael; Szymon Wilk (1999), *Determinants of Long-Term Economic Development: An Empirical Cross-Country Study Involving Rough Sets Theory and Rule Induction* (Transition Economics Series No. 11), Vienna: Institute for Advanced Studies.

Ohlin, Bertil (1933), *Interregional and International Trade*, Cambridge.

Olson, Mancur (1982), *The Rise and Decline of Nations: Economic Growth, Stagflation and Social Rigidities*, New Haven: Yale University Press.

Onyemelukwe, Clement C. (2005), *The Science of Economic Development and Growth: The Theory of Factor Proportions,* Armonk N.Y.: Sharpe.

Osella, Caroline (2011), "The Gulf and its Indian Entanglements", *The Middle East in London*, April-May.

Ouled Belayachi, Karim; Jamal Ibrahim Haidar (2008), "Competitiveness from Innovation, not Inheritance", in: Bruce Ross-Larson (ed.), *Celebrating Reforms 2008 – 16 New Case Studies*, The World Bank. http://www.doingbusiness.org/documents/CaseStudies/SaudiArabia_CS2008.pdf [16 January 2014].

Ouroussoff, Nicolai (2010), "Planning a Sustainable City in the Desert", *The New York Times*, September 25.

Owen, Roger; Şevket Pamuk (1999), *A History of Middle East Economies in the Twentieth Century*, Cambridge MA: Harvard University Press.

Owtram, Francis (2004), *A Modern History of Oman. Formation of the State since 1920*, London; New York: I.B. Tauris.

Oxford Analytica, "Saudis Renew Search for Food Security", *Gulf News*, April 23, 2009. http://gulfnews.com/business/opinion/saudis-renew-search-for-food-security-1.65122 [22 December 2013].

Oxford Business Group (2012), "The Report: Bahrain 2012", http://www.oxfordbusinessgroup.com/shop/country/Bahrain [7 January 2014].

Palley, Thomas I. (2003), "Lifting the Natural Curse", *Foreign Service Journal*, 80, 12, pp. 54-61.

Palma, Gabriel (1978), "Dependency: A Formal Theory of Underdevelopment or a Methodology for the Analysis of Concrete Situations of Underdevelopment", *World Development*, 6, pp. 881-924.

Pawelka, Peter (1993), *Der Vordere Orient und die internationale Politik*, Stuttgart; Berlin; Köln: W. Kohlhammer.

Peterson, J.E. (2012), "Oman Succesion Unlikely to Jeopardise Security", *Oxford Analytica*, 8. November.

Peterson, J.E. (2013), "The Solitary Sultan and the Construction of the New Oman", in: Abbas Kadhim (ed.), *Governance in the Middle East and North Africa: A Handbook*, London: Routledge.

Pfeifer, Karen (2002), "Kuwait's Economic Quandary", *Middle East Report*, 223, pp. 10-13.

Philipp, Thomas (1990), "The Idea of Islamic Economics", *Die Welt des Islams*, New Series, 30, 1 /4, 1990, pp. 117-139.

Piketty, Thomas (2014), *Capital in the Twenty-First Century*, Harvard MA: Harvard University Press.

Pourgerami, Abbas (1988), "The Political Economy of Development: A Cross-National Causality Test of Development-Democracy-Growth Hypothesis", *Public Choice*, 58, 2, pp. 123-141.

Prados, A.B.; C. M. Blanchard (2004), "Saudi Arabia: Terrorist Financing Issues", December 8, 2004. http://www.fas.org/irp/crs/RL32499.pdf [15 January 2014].

Preston, P. W. (1996), *Development Theory. An Introduction*, Oxford; Cambridge, MA: Blackwell.

Prokop, Michaela (2005), "The War of Ideas: Education in Saudi Arabia", in: Aarts/Nonneman (ed.), *Saudi Arabia in the Balance: Political Economy, Society, Foreign Affairs*, London; Hurst 2005, pp. 57-82.

Quinn, Dennis P.; John T. Woolley (2001), "Democracy and National Economic Performance: The Preference for Stability", *American Journal of Political Science*, 45, 3, pp. 634-657.

Rabi, Uzi (2011), *The Emergence of States in a Tribal Society. Oman Under Sa'id bin Taymur, 1932-1970*, Eastbourne: Sussex Academic Press.

Raffer, Kunibert; Mohamed Abdel Rahim Mohamed Salih eds. (1992), *The Least Developed and the Oil-Rich Arab Countries. Dependence, Interdependence or Patronage*, Houndmills; London: Macmillan Press.

Raffer, Kunibert; Hans W. Singer (2001), *The Economic North-South Divide. Six Decades of Unequal Development*, Cheltenham; Northampton: Edward Elgar.

Ragab, Ibrahim A.(1980), "Islam and Development", *World Development*, 8, pp. 513-521.

Ragin, Charles C (1987), *The Comparative Method. Moving beyond Qualitative and Quantitative Strategies*, Berkeley/Los Angeles/London: Univ. of California Press.

Ragin, Charles C. (2000), *Fuzzy-Set Social Science*, University Chicago Press.

Ragin, Charles; John Sonnett (2004), "Between Complexitiy and Parsimony: Limited Diversity, Counterfactual Cases, and Comparative Analysis", in Sabine Kropp; Michael Minenberg (eds), *Vergleichen in der Politikwissenschaft*, Wiesbaden: Verlag für Sozialwissenschaften, pp. 147-159.

Ragin Charles C. (2006), "The Limitations of Net-Effects Thinking", in: Rihoux, Benoît; Heike Grimm, (eds.), *Innovative Comparative Methods for Policy Analysis. Beyond the Quantitative-Qualitative Divide*, Springer, pp. 13-41.

Ragin, Charles C. (2008), "Measurement Versus Calibration: A Set-Theoretic Approach", in Janet Box-Steffensmeier; David Collier; Henry Brady (eds.), *Oxford Handbooks of Political Methodology*, Oxford: Oxford University Press 2008.

Ragin, Charles; Sean Davey (2012), *fs/QCA [Computer Programme]*, Version 2.5. Irvine, CA: University of California.

Ram, Rati (1986) "Government Size and Economic Growth: A New Framework and Some Evidence from Cross-Section and Time-Series Data", *The American Economic Review*, 76, 1, pp. 191-203.

Ramady, Mohamed A. (2013), *Political, Economic and Financial Country Risk. Analysis of the Gulf Cooperation Council*, Cham; Heidelberg; London: Springer.

Ramady, Mohamed A. (2010), *The Saudi Arabian Economy. Policies, Achievements and Challenges*, Cham; Heidelberg; London: Springer 2nd edition.

Ramady, Mohamed and W. Mahdi (forthcoming), *OPEC in a shale- oil world: Where to next?*, Heidelberg: Springer.

Al-Rasheed, Madawi (2002), *A History of Saudi-Arabia*, Cambridge: Cambridge University Press.

Rathmell, Andrew; Kirsten Schulze (2000), "Political Reform in the Gulf: The Case of Qatar", *Middle Eastern Studies*, 36, 4, pp. 47-62.

Reuters, "Update 1-No consensus on Morocco, Jordan joining GCC", Reuters, November 30, 2011, http://www.reuters.com/article/2011/11/30/gulf-membership-idAFL5E7MU2YL20111130 [2013-12-22].

Ricardo, David (1962), "On the Principles of Political Economy and Taxation", in: Piero Sraffa (ed.), *The Works and Correspondence of David Ricardo*, 1, Cambridge.

Richards, Alan; John Waterbury (2008), *A Political Economy of the Middle East*, Boulder, CO: Westview Press.

Riese, Hajo (1983), "Geldökonomie, Keynes und die Anderen. Kritik der monetären Grundlagen der Orthodoxie", in: P. de Gijsel a. o. (eds.): *Ökonomie und Gesellschaft. Jahrbuch 1: Die Neoklassik und ihre Herausforderungen,* Frankfurt/Main.: Campus, pp. 103-160.

Rihoux, Benoît; Charles C. Ragin (2008), *Configurational Comparative Methods. Qualitative Comparative Analysis (QCA) and Related Techniques*, (Applied Social Research Methods Series, 51), Los Angeles; London a. o.: Sage.

Robinson, James A.; Ragnar Torvik; Thierry Verdier (2006), "Political Foundations of the Resource Curse", *Journal of Development Economics*, 79, pp. 447-468.

Rodinson, Maxime (1966), *Islam et capitalisme*, Paris: Éditions du Seuil.

Rodinson, Maxime (2002), *Muhammad: Prophet of Islam*, London; New York: I.B. Tauris.

Rodríguez, Francisco; Dani Rodrik (2000), "Trade Policy and Economic Growth: A Skeptic's Guide to the Cross-National Evidence", *NBER Macroeconomics Annual*, 15, pp. 261-325.

Romer, Paul M. (1986), "Increasing Returns and Long-Run Growth", *Journal of Political Economy,* 94, pp. 1002-1035.

Romer, Paul M. (1990), "Endogenous Technological Change", *Journal of Political Economy*", 98, pp. 71-102.

Romer, Paul M. (1994) "The Origins of Endogenous Growth", *The Journal of Economic Perspectives*, 8, 1, Winter, 1994, pp. 3-22.

Ros, Jaime (2001), *Development Theory and the Economics of Growth* (Development and Inequality in the Market Economy), Ann Arbor: The University of Michigan Press.

Rosenstein-Rodan, Paul N. (1943), "Problems of Industrialization of Eastern and South-Eastern Europe", *Economic Journal,* 53, 210/211, p. 202-11.

Rosenstein-Rodan, Paul N. (1944), "The International Development of Economically Backward Areas", *International Affairs,* 20/2, pp. 157-165.

Ross, Michael (2001), "Does Oil Hinder Democracy?", *World Politics*, 53, 3, pp. 325-361.

Ross, Michael (2008), "Oil, Islam, and Women", *American Political Science Review*, 102, 1, pp. 107-123.

Rostow, Walt Whitman (1960), *The Stages of Economic Growth. A Non-Communist Manifesto*, Cambridge: Cambridge University Press.

Rothlauf, Jürgen (2006), *Interkulturelles Management: mit Beispielen aus Vietnam, China, Japan, Russland und den Golfstaaten* (WiSo-Lehr- und Handbücher), München; Wien: Oldenburg.

Sachs, Jeffrey D.; Andrew M. Warner (1995a/1997), "Natural Resource Abundance and Economic Growth", *Working Paper* 5398, Cambridge, MA: National Bureau of Economic Research and Harvard University.

Sachs, Jeffrey D.; Andrew M. Warner (1997), "Fundamental Sources of Long-Run Growth", *The American Economic Review*, 87, 2, Papers and Proceedings of the Hundred and Fourth Annual Meeting of the American Economic Association, pp. 184-188.

Sachs, Jeffrey D.; Andrew M. Warner (1999), "The Big Push, Natural Resource Booms and Growth", *Journal of Development Economics*, 59, pp. 43-76.

Sachs, Jeffrey D.; Francisco Rodriguez (1999), "Why Do Resource-Abundant Economies Grow More Slowly?", *Journal for Economic Growth*, 4, pp. 277- 303.

Sachs, Jeffrey D.; Andrew M. Warner (2001), "Natural Resources and Economic Development: The Curse of Natural Resources", *European Economic Review*, 45, pp. 827-838.

Sachs, Jeffrey D.; Macartan Humphreys; Joseph E. Stiglitz (2007), *Escaping the Resource Curse,* New York: Columbia University Press.

Sadik, Muhammad T., William P. Snavely (1972), *Bahrain, Qatar, and the United Arab Emirates. Colonial Past, Present Problems, and Future Prospects*, Lexington; Toronto; London: Lexington Books.

Sadowski, Yahya (1997), "The End of the Counterrevolution?: The Politics of Economic Adjustment in Kuwait", *Middle East Report*, 204, The Arabian Peninsula, pp. 7-11+26.

Schacht, Joseph (1965), *An Introduction to Islamic Law*, Oxford: Clarendon Press 2ed (reprint of 1998).

Schmid, Claudia (1991), *Das Konzept des Rentier-Staates: Ein sozialwissenschaftliches Paradigma zur Analyse von Entwicklungsgesellschaften und seine Bedeutung für den Vorderen Orient* (Demokratie und Entwicklung 29), Münster: Lit-Verlag.

Schmid, Claudia (1997), "Rente und Rentier-Staat. Ein Beitrag zur Theoriengeschichte", in: Beck, Martin; Andreas Boeckh, Peter Pawelka (eds.), *Staat, Markt und Rente in der internationalen Politik,* Opladen: Budrich, pp. 28-50.

Schneider, Carsten Q., Claudius Wagemann (2007), *Qualitative Comparative Analysis (QCA) and Fuzzy Sets. Ein Lehrbuch für Anwender und jene, die es werden wollen.* Mit einem Vorwort von Charles Ragin. Opladen; Farmington Hills: Barbara Budrich.

Schubert, Sven (2009), "Kaum Chancen für Einheitswährung am Golf bis 2010", *Credit Suisse Research Focus*, March 23.

Schumpeter, Joseph A. (1942), *Capitalism, Socialism, and Democracy*, New York: Harper & Row.

Senghaas, Dieter (1999), "Friedrich List (1789-1846). Binnenmarktorientierung als Basis wirtschaftlicher Entwicklung", *Entwicklung und Zusammenarbeit*, 6, pp. 164-168.

Senghaas, Dieter ed. (1974), *Peripherer Kapitalismus. Analysen über Abhängigkeit und Unterentwicklung*, Frankfurt: Suhrkamp Verlag.

Seznec, Jean-François; Mimi Kirk (2010), *Industrialization in the Gulf: A Socioeconomic Revolution*, New York: Routledge.

Shabana, Ramy; Sonia Salah (2013), "The New Kuwait Foreign Investment Law", *AL Tamimi & CO Law Update*, September.

Sharpley, R. (2002), "The Challenges of Economic Diversification through Tourism: The Case of Abu Dhabi", *International Journal of Tourism Research*, 4, 3, pp. 221-35.

Shihab, Mohamed (2001), "Economic Development in the UAE", in: Ibrahim al Abed; Peter Hellyer (eds.), *United Arab Emirates. A new perspective*, London: Trident Press, pp. 249-259.

Shochat, Sharon (2008), "The Gulf Cooperation Council Economies. Diversification and Reform. An Introduction", *Kuwait Programme on Development, Governance and Globalisation in the Gulf States/ LSE*, Background Paper, London: LSE.

Simmons, Matthew R. (2005), *Twilight in the Desert: The Coming Saudi Oil Shock and the World Economy*, Hoboken, NJ: Wiley.

Singer, Hans W. (1999), "How Relevant is Keynesianism Today for Understanding Problems of Development?", in: Elaine Petitat-Côté (ed.), *The EADI Story. Networking over the Years 1975-2000*, Genf: EADII, pp. 145-156.

Smith, Benjamin (2005), "Oil Wealth and Regime Change", Conference Paper *Transforming Authoritarian Rentier Economies*, Bonn, September 22-23.

Solow, Robert M. (1956), "A Contribution to the Theory of Economic Growth", *The Quarterly Journal of Economics*", 70, 1, pp. 65-94.

Speece, Mark (1989), "Aspects of Economic Dualism in Oman, 1830-1930", *International Journal of Middle East Studies*, 21, 4, pp. 495-515.

StataCorp. (2009), *Stata Statistical Software: Release 11*. College Station, TX: StataCorp LP.

Stauffer, Thomas R.; Frank H. Lennox (1984), "Accounting for 'Wasting Assets'. Income Measurement for Oil and Mineral-Exporting Rentier States", *OFID Pamphlet Series* 25.

Stauffer, Thomas (1987), "Income Measurement in Arab States", in: Beblawi/Luciani (eds.), *The Rentier State (Nation, State and Integration in the Arab World Vol. II)*, London; New York; Sydney: Croom Helm, pp. 22-48.

Steinberg, Guido (2001), "The Shiites in the Eastern Province of Saudi Arabia (al-Aḥsāʾ). 1913-1953", in: Brunner, Rainer; Werner Ende (eds.), *The Twelver Shia in Modern Times. Religous Culture & Political History* (Social, Economic and Political Studies of the Middle East and Asia, 71), Leiden; Boston; Köln: Brill, pp. 236-254.

Steinberg, Guido (2005), "The Wahhabi Ulama and the Saudi State: 1745 to the Present", in: Aarts/ Nonneman (eds.), *Saudi Arabia in the Balance: Political Economy, Society, Foreign Affairs*, London: Hurst, pp. 11-34.

Stiglitz, Joseph (2005), "Finance for Development", in: Ayogu, Melvin; Don Ross (eds.), *Development Dilemmas. The Methods and Political Ethics of Growth Policy* (Routledge studies in development economics 41), London; New York: Routledge, pp. 15-29.

Sturm, Michael; Nikolaus Siegfried (2005), "Regional Monetary Integration in the Member States of the Gulf Cooperation Council", *European Central Bank: Occasional Paper Series,* 31.

Sukidi (2006), "Max Weber's Remarks on Islam: The Protestant Ethic among Muslim Puritans", *Islam and Muslim-Christian Relations*, 17, 2, pp. 195-205.

Al-Tarrah, Ali (2007), "Family in the Kinship State", in: Alsharekh, *The Gulf Family. Kinship Policies and Modernity* (SOAS Middle East Issues), London; Berkeley, Beirut: Saqi, pp. 119-124.

Tétreault, Mary Ann; Mohammed al-Ghanim (2009), "The Day After "Victory": Kuwait's 2009 Election and the Contentious Present", *MERIP Online*, July 8.

Thirlwall, Anthony P. (2003), *Growth and Development with Special Reference to Developing Economies*, Palgrave[7].

Thompson, Mark (2015), "Self-Motivation, Career Aspirations and Work Responsibilities: The Perspective of Saudi Male Undergraduates at King Fahd University of Petroleum and Minerals", in: Annika Kropf and Mohamed A. Ramady (eds.), *Employment and Career Motivation in the Arab Gulf States: The Rentier Mentality Revisited*, Berlin: Gerlach Press (forthcoming).

Tibi, Bassam (1985), *Der Islam und das Problem der kulturellen Bewältigung sozialen Wandels*, Frankfurt /Main.

Tibi, Bassam (1991), *Die Krise des modernen Islams: eine vorindustrielle Kultur im wissenschaftlich-technischen Zeitalter. Mit e. Essay: Islamischer Fundamentalismus als Antwort auf die doppelte Krise,* Frankfurt am Main: Suhrkamp[2].

Toumi, Habib, "Gulf countries welcome bid from Joran, Morocco to join the GCC bloc", *Gulf News*, May 11, 2011, http://gulfnews.com/news/region/jordan/gulf-countries-welcome-bid-fromjordan-morocco-to-join-gcc-bloc-1.806216 [22 December 2013]

Townsend, John (1980), "Problems Confronting the Establishment of a Heavy Industrial Base in the Arab Gulf", in: Tim Niblock (ed.), *Social and Economic Development in the Arab Gulf*, London: Croom Helm, pp. 95-105.

Turner, Bryan S. (1998), *Weber and Islam: A Critical Study*, London: Routledge.

UAE (United Arab Emirates) Cabinet (2010), *UAE Vision 2021: United in Ambition and Determination*, Abu Dhabi: Government of United Arab Emirates.

U.S. Energy Information Administration (2013), Bahrain, http://www.eia.gov/countries/country-data.cfm?fips=ba [7 January 2014].

Vicente, Pedro C. (2006), "A Theory of Corruption, Political Exclusion, and Windfalls", in: Gradstein; Konrad (eds.), *Institutions and Norms in Economic Development*, London; Cambridge, MA.: The MIT Press, pp. 111-139.

Voigt, Stefan (2003), "Neue Institutionenökonomie als kulturelle Ökonomik", in: Blümle/Goldschmidt/ Klump a. o. (eds.), *Perspektiven einer kulturellen Ökonomik (Kulturelle Ökonomik Bd. 1)*, Münster: LIT Verlag, pp. 411-426.

Waterbury, John (1970), *The Commander of the Faithful. The Moroccan Political Elite. A Study in Segmented Politics*, New York: Columbia University Press.

Weber, Max (2006), *Die protestantische Ethik und der Geist des Kapitalismus, Vollständige Ausgabe* introduced by Dirk Kaesler (ed.), München: C. H. Beck[2].

Whelan, John (1987), "Oman's Development Strategy", in: B.R. Pridham (ed.), *Oman: Economic, Social and Strategic Developments*, pp. 134-144.

Wilkinson, John C. (2010), *Ibadism. Origins and Early Development in Oman*, Oxford: Oxford University Press.

Willoughby, John (2008), "Let a Thousand Models Bloom: Forging Alliances with Western Universities and the Making of the New Higher Educational System in the Gulf", *Working Paper* No. 2008-01. Washington, DC: Department of Economics, American Univeristy.

Wilson, Peter W. (1991), *A Question of Interest: The Paralysis of Saudi Banking*, Boulder, CO: Westview 1991.
Wilson, Peter W.; Douglas Graham (1994), *Saudi Arabia: The Coming Storm*, New York: M.E. Sharpe.
Wilson, Rodney (1982), "Economic Change and Re-Interpretation of Islamic Social Values", *Bulletin of the British Society for Middle Eastern Studies*, 9, 2, pp. 107-113.
Wilson, Rodney (1982a), "The Evolution of the Saudi Banking System and its Relationship with Bahrain", in: Tim Niblock, *State, Society, and Economy in Saudi Arabia*, Beckenham: Croom Helm, pp. 278-300.
Winckler, Onn (2009), *Arab Political Demography. Population Growth, Labor Migration and Natalist Policies* (Demographic Developments and Socioeconomic Policies in the Middle East and North Africa), Sussex Academic Press.
Wittman, Donald (1989), "Why Democracies Produce Efficient Results", *Journal of Political Economy*, 97, pp. 1395-1424.
Wößmann, Ludger (2004), "Specifying Human Capital", in: George, Donald A.R.; Les Oxley; Kenneth I. Carlaw (eds.), *Surveys in Economic Growth. Theory and Empirics*, Malden, MA; Oxford; Victoria: Blackwell Publishing, pp. 13-44.
WTO (2005), "The Report of the Working Party on the Accession of the Kingdom of Saudi Arabia to the World Trade Organization", http://www.wto.org/english/thewto_e/acc_e/completeacc_e.htm#sau [15 January 2014].
Younger, Stephen D. (1992), "Aid and the Dutch Disease: Macroeconomic Management When Everybody Loves You", *World Development*, 20, 11, pp. 1587-1597.
Zahlan, Rosemarie Said (1998), *The Making of the Modern Gulf States: Kuwait, Bahrain, Qatar, the United Arab Emirates and Oman*, Reading: Ithaca Press (revised and updated edition).
Zaman, Raguibuz (1980), "Policy Implications of Introducing Zaka into Bangladesh and Saudi Arabia", in: Zaman, Raguibuz (ed.), *Some Aspects of the Economics of Zakah*, Plainfield, IN, p. 80.
Zogby, James J. (2007), "Shedding Light on the Gulf's Middle Class", *The McKinsey Quarterly: Sepcial Edition: Reappraising the Gulf States*, pp. 31-37.

Newspapers

Arabian News, Gulf News, Khaleej Times, Kuwait News, MEED Middle East Economic Digest, MERIP Online, The Guardian, The New York Times, The Telegraph, The Natinal (UAE)

Interviews Quoted

Brij Bali, Head of Student Service, Abu Dhabi Higher Colleges of Technology, interviewd by author, Abu Dhabi, February 2009.

Jane Brystol-Rhys, Professor of Anthropology, Zayed University, Abu Dhabi, February 2009.

Monika Fatima Mühlböck, Professor of Islamic Studies, Vienna, interviewed by author, Vienna, January 2010.

Various conversations with students at Abu Dhabi Higher Colleges of Technology, with Rola Dashti, former MP in Kuwait, Prof. Mohamed A. Ramady, KFUPM, Dhahran, Saudi Arabia, Prof. Kunibert Raffer, Department of Macroeconomics, University of Vienna, Tanya Cariina Hsu, advisor to the Saudi Royal Family, and with many other consultants or expatriates working in the GCC states.

Websites (selection)

Arabian Business
www.arabianbusiness.com

Arab Monetary Fund
www.amf.org.ae

BBC
http://www.bbc.co.uk/

British Petroleum
www.bp.com

CIA Factbook
www.cia.gov

Compasss – Small-N Systematic cross-case analysis
http://www.compasss.org/

Corruption Perception Index (by Transparency International)
http://www.transparency.org/policy_research/surveys_indices/cpi

Doing Business (World Bank group)
www.doingbusiness.org/

DPA
http://www.dpa.de/

Economic and Social Commission for Western Asia
http://www.escwa.un.org/

Economist Intelligence Unit
www.eiu.com

Energy Information Administration
www.eia.doe.gov

European Union Extgernal Action
http://www.eeas.europa.eu/

Federation of American Scientists
www.fas.org

Food and Agriculture Organization of the United Nations/Statistics - FAOSTAT
http://faostat.fao.org/

Freedom House
www.freedomhouse.org

Global Footprint Network
http://www.footprintnetwork.org/en/index.php/GFN/

Gulf Cooperation Council
http://www.gcc-sg.org/

Gulf News
http://gulfnews.com/

Heritage Foundation
http://www.heritage.org/

Human Rights Watch
http://www.hrw.org/

International Fund for Agricultural Development (IFAD)
www.ifad.org

International Monetary Fund
www.imf.org

McKinsey&Company
http://www.mckinsey.com/

Middle East Business Intelligence (MEED)
http://www.meed.com/

Middle East Research and Information Project
http://www.merip.org/

Moody's
http://www.moodys.com/

NY Law Global
http://www.law.nyu.edu/global/index.htm

OECD
www.oecd.org

OPEC
http://www.opec.org/opec_web/en/

Oxford Analytica
https://www.oxan.com/

Oxford Business Group
http://www.oxfordbusinessgroup.com/

Penn World Table
http://pwt.econ.upenn.edu/

Reuters
http://faostat.fao.org/

Reuters
www.reuters.com

Revenue Watch Institute
www.revenuewatch.org

SAMA
http://www.sama.gov.sa/sites/SAMAEN/Pages/Home.aspx

SAMIRAD – Saudi Arabia Market Information Resource and Directory
http://www.saudinf.com/index.htm

Science Direct
www.sciencedirect.com

Standard & Poor's
http://www.standardandpoors.com/home/en/eu

Sovereign Wealth Fund Institute
http://www.swfinstitute.org/

Stockholm International Peace Research Institute
http://www.sipri.org/

United Nationas Conference on Trade and Development
www.unctad.org

UNDP
www.undo.org

UNESCO
http://www.unesco.org/new/en/unesco/

US Energy Informatin Administration
http://www.eia.gov/

Worldbank
www.worldbank.org

World Value Survey
http://www.worldvaluessurvey.org/